This Book
Belongs To
Mary Elizabeth Turner

HORSE BEHAVIOR

NOYES SERIES
IN
ANIMAL BEHAVIOR, ECOLOGY,
CONSERVATION AND MANAGEMENT

A series of professional and reference books in ethology devoted to the better understanding of animal behavior, ecology, conservation, and management.

HORSE BEHAVIOR

The Behavioral Traits and Adaptations of Domestic and Wild Horses, Including Ponies

by

GEORGE H. WARING

Southern Illinois University
Carbondale, Illinois

np **NOYES PUBLICATIONS**
Park Ridge, New Jersey, USA

Library of Congress Catalog Card Number: 82-19083
ISBN: 0-8155-0927-8
Printed in the United States

Published in the United States of America by
Noyes Publications
Mill Road, Park Ridge, New Jersey 07656

10 9 8 7 6 5 4 3 2 1

Library of Congress Cataloging in Publication Data

Waring, George H.
Horse behavior.

(Animal behavior, ecology, conservation, and
management)
Bibliography: p.
Includes index.
1. Horses--Behavior. I. Title. II. Series.
SF281.W37 1983   599.72'5   82-19083
ISBN 0-8155-0927-8

This book is dedicated to my Family
        and to Equine Researchers worldwide,
to Equids past and present,
        and to the Almighty who created us all.

Preface

To the avid horseman, this book will provide a comprehensive overview and reference to scientific studies of horse behavior. Data from studies throughout the world are included. Sources of information are cited within the text and are listed in the Bibliography. To veterinarians and students of veterinary science, the book will provide a baseline of typical horse traits and contrast those with abnormalities encountered in equine medicine. To readers interested in animal husbandry, the content of the book will provide ethological guidance for successful management, handling, and production. And to ethologists and to students of natural history, the book will provide insight into the behavioral biology and adaptations of a truly fascinating species–*Equus caballus.*

The material considers the horse, including ponies, under both domesticated and feral conditions. No attempt is made to also review the traits of the other equine species. Technical terms pertaining to behavior are clarified within the text. When using the volume as a reference, the Index and Table of Contents will be especially helpful. Figure 1.3 should prove useful when clarification of anatomical terminology is needed.

Acknowledgements are due to the following who assisted in reviewing the manuscript, in aiding my research, or in giving permission to use illustrations: Ronald D. Carr, O.J. Ginther, Gertrude Hendrix, David M. Lane, Werner Leidl, Susan Marinier, Kam Matray, Eugene Morgan, Delyte W. Morris, Patricia A. Noden, B.W. Pickett, N.O. Rasbech, James R. Rooney, Peter D. Rossdale, Stephanie J. Tyler, Gail S. Van Asten, and Ann-Meredith Waring.

Permission for the reproduction of figures was graciously given by: *American Journal of Veterinary Research,* U.S.A.; Baillière Tindall, England; *Journal of Animal Science,* U.S.A.; *Journal of Reproduction and Fertility,* England; *Modern Veterinary Practice,* U.S.A.; Société de Biologie, France; and Springer Verlag, U.S.A.

Drawings were made by the skilled hand of Gail S. Van Asten. Photographic credits go to Peter D. Rossdale, Philip Malkas, Ronald R. Keiper, and to my camera. Daniel Klem, Jr. and Albert Kipa assisted with translations. The staff of Noyes, especially Alice W. Pucknat, contributed skilled and timely aid; I gratefully acknowledge the role George Narita had in promoting this book from its inception.

To all the above and for the encouragement given by fellow ethologists, I give my sincere thanks. My gratitude is also expressed to my wife Ann-Meredith; to my children Sari, Houstoun, and Heidi; and to my parents Houstoun and Irene for their interest in my work and their devotion.

Southern Illinois University   George H. Waring
   at Carbondale
August 1982

Contents

x   Contents

Mutual grooming

spatial rel.

May 18

Part I

Introduction

1

Ancestry of the Horse

We experience horses and their behavioral traits at a relatively brief point in time; yet horses and other equids have not always been the way we see them today. Many changes have occurred, spanning millions of years. The changes that are recorded in the geological record appear to have been sporadic, probably in response to changing environments and as a result of changing genetic composition. A view of the ancestry of the horse provides a foundation to understand the ethology of the subject of this book–the domestic horse.

A member of the family Equidae, the horse is placed with other recent equids into the genus *Equus*. The domestic horse, *Equus caballus,* is one of the several living equid species. These include the Przewalski horse, African ass, Asian ass, and the zebras (Table 1.1).

Among the living equids, the domestic horse is most like the Przewalski horse. Chromosomal studies reveal many similarities; nevertheless, consistent differences also occur (Ryder et al. 1978). Domestic horses have a diploid (2n) chromosome number of 64, whereas Przewalski horses have 66 chromosomes. Although such a disparity may indicate they are each distinct species (Benirschke and Malouf 1967), they could be part of a single species exhibiting chromosomal polymorphism, as occurs in several mammalian species from mice to some large artiodactyls (Epstein 1971) and even the Asian ass (Ryder 1977). Fusing two Przewalski chromosome pairs together would account for the reduced number of chromosomes in domestic horses. Crosses of Przewalski and domestic horses (each having a cytogenetic fundamental number of 92) produce fertile offspring which have body cells with a diploid chromosome compliment of 65. Blood group and serum protein studies also indicate a similarity between Przewalski and domestic horses (Podliachouk and Kaminski 1971). Unfortunately, some domestic horse genes may occur in some Przewalski stock commonly available for research as a result of an early crossbreeding (Dolan 1962). Free-roaming Przewalski horses may now be extinct; survivors of the species are found in zoos.

Most equid species are known only from fossil remains. Numerous extinct species have been described. Fossil materials from Eocene deposits up to recent times give an excellent overview of equid evolution, especially in North America. It was not orthogonal or straightline evolution, as we sometimes simplify in our mind. For example, when viewed as a whole, there was no constant and overall increase in body size, the legs did not sequentially lengthen, and the feet did not steadily change from four toes to three and finally one. Some lines decreased body size and limb length, while others retained body and limb characteristics relatively unchanged for long periods. Trends varied. Numerous combinations are found. In one genus, for example, certain characteristic changes would be present that would not occur in other evolutionary lines. There were numerous branchings to the family tree, as illustrated in Figure 1.1, and only certain genetic lines survived the rigors of the changing environment over the ages.

**Table 1.1: Classification of the Horse and Related Species
of Living Equids**
Listed Sequentially by Diploid Chromosome Number (in parentheses)

Class Mammalia
  Order Perrisodactyla
    Family Equidae
      Genus Equus
        Species

(66) *Equus przewalskii* (Przewalski horse, Wildpferd, cheval de Przewalski, caballo salvaje Przewalski)
(64) *Equus caballus* (domestic horse, Hauspferd, cheval, caballo)
(62) *Equus asinus* (African ass, donkey, Esel, anès, burro)
      E. asinus africanus (Nubian wild ass)
      E. asinus somalicus (Somali wild ass)
(56) *Equus hemionus* (Asian ass, Halbesel, hémione, asno asiático)
      E. hemionus hemionus (Mongolian kulan)
      E. hemionus onager (onager)
      E. hemionus khur (Indian wild ass)
      E. heminous kiang (kiang)
(46) *Equus grevyi* (Grevy's zebra, Grévyzebra, zèbre de Grévy, cebra de Grevy)
(44) *Equus burchelli* (Burchell's zebra, Steppenzebra, zèbre de steppe, cebra de Burchell)
      E. burchelli antiquorum (Chapman's zebra)
      E. burchelli boehmi (Grant's zebra)
      E. burchelli selousi (Selous's zebra)
(32) *Equus zebra* (mountain zebra, Bergzebra, zèbre de montagne, cebra de montaña)
      E. zebra zebra (Cape mountain zebra)
      E. zebra hartmannae (Hartmann's zebra)

When we consider just those ancestral forms that led directly to the present equids, we find that over about sixty million years horse evolution went from the dog-like *Hyracotherium,* commonly called eohippus, with four toes on the forelegs and three on the hind, to the genus *Equus,* with a single digit supporting each leg. Simpson (1951) has carefully outlined this evolutionary history, the basis of the following summary.

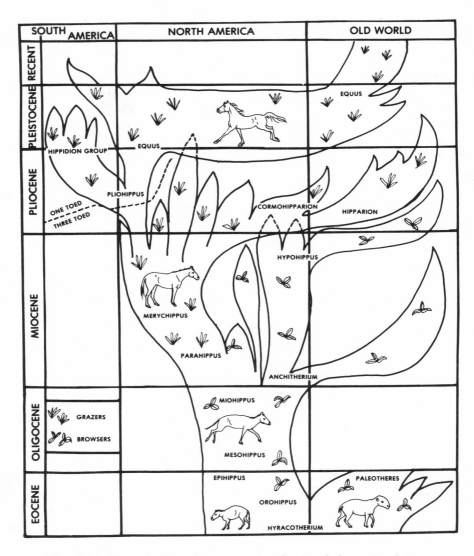

Figure 1.1: The main lines of horse ancestry, showing phylogenetic relationships. (Modified from Simpson 1951)

There were different species of eohippus, and they were widespread in the northern hemisphere. Yet, judging from tooth characteristics, they all were browsers eating succulent leaves and lesser amounts of soft seeds and small fruits. These animals varied greatly in size from approximately 25 to 50 centimeters (10-20 inches) high at the shoulders, and some species were probably eight times heavier than adults of other eohippus species. They had arched, flexible backs, and their tails were long and stout. Each of the toes ended in a separate small hoof. The body weight was carried not on the hooves but primarily on a dog-like pad. The lower leg was not vertical in the standing position as we associate with modern equids; Sondaar (1968, 1969) points out that the metapodials of early equids had an obvious slope while in a resting stance (Figure 1.2). The limb construction and the flexible back suggest changes in locomotor patterns have definitely occurred between these ancient forms and the modern equids. Compared to the possible phenacodontid condylarth ancestors of the Paleocene, *Hyracotherium* species show increased specialization for running (Radinsky 1966).

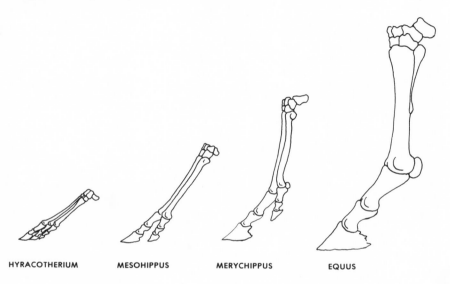

HYRACOTHERIUM   MESOHIPPUS   MERYCHIPPUS   EQUUS

Figure 1.2: Evolution of the equine foot. Side view of forefeet in resting position. Drawn to scale. (Adapted from Simpson 1951 and Sondaar 1969.)

The skull of eohippus was only slightly proportional to that of a horse. The brain was small and so primitive that it resembled the most primitive mammal brains. The dentition consisting of 44 teeth was beginning to show a trend where the front set of teeth was used for nipping and picking up food and a separate back set was used for crushing and grinding food. The jaw musculature suggests increased specialization for lateral jaw movement typical of herbivores (Radinsky 1966). The horse system of manipulating the food with the tongue was probably also appearing in eohippus.

In the early Eocene equids designated as *Hyracotherium*, all the premolars were unlike the molars and the crests on the cheek teeth were not well developed. Middle Eocene equids, *Orohippus*, and the late Eocene equids, *Epihippus*, retained the low-crowned teeth of eohippus but showed progressive development of premolars with molar-like patterns (molarization) and the development of crested or ridged cheek teeth (lophiodonty). These later equids thus had more strictly herbivorous dentition and more effective teeth for browsing.

The genus *Mesohippus* contains the earliest equids known to have had only three toes on the front feet. Such animals probably appeared very much like small horses as they roamed North America in early and middle Oligocene. An equine muzzle was probably present, but the eye was not yet as far back as in recent horses. The brain case was now swollen, and fossil brain casts show the cerebral hemispheres had become relatively much larger and the surfaces had become convoluted with a series of folds and grooves. The brain was similar in type to a modern ungulate brain. The initial development of the characteristic equine intelligence thus took place during the transition from Eocene to Oligocene and not with the origin of the family. The *Mesohippus* brain was, nevertheless, distinctly more primitive than in later and more recent equids.

The teeth of *Mesohippus* species were low-crowned and still fitted for browsing, not grazing. The second to fourth premolars were very much like molars in pattern and thus the cheek teeth were a set of crushing and grinding teeth all similar in appearance.

The legs of *Mesohippus* were long and slender, and the animals had three fully functional toes on each foot with a pad between and behind them to support the main weight of the body. At rest the metacarpals made an angle of about 50° with the horizontal plane (Figure 1.2) unlike later equids whose forelegs became more vertical in the resting position (Sondaar 1969).

In features of the foot and in many other characteristics, the *Miohippus* species of mid and late Oligocene were similar to *Mesohippus,* but with *Miohippus* the metatarsal (cannon bone) of the third or middle toe came into contact with not only the ankle bone called the ectocuneiform but also with the cuboid, achieving greater stability in the hock. The three-toed feet of these animals were of advantage in soft soil of forests or along river banks where they likely fed on mature leaves of trees and bushes. The musculature and action of the foot allowed these animals to pull their toes together as the foot was lifted to ease removal of the foot from mud or soft sand.

Miohippus existed into the early Miocene, and there its fossils intergrade with several different descendant groups. Most of these groups diversified further as three-toed browsers. Some migrated from North America to the Old World where the various descendents of eohippus had long become extinct. These browsers also eventually vanished (see Figure 1.1).

One line of development from *Miohippus* did continue, however, in North America. These equids were beginning to eat grass, and their teeth and digestive system continued to change to enable them to utilize the abrasive, high fiberous foods. Grasses were then becoming common in the cooling and drying environment, replacing tropical flora. These equids exploited this new niche. The excellent fossil record shows gradual changes from *Parahippus* of early Miocene to the mid and upper Miocene descendants placed by paleontologists into the genus *Merychippus.* Among the tooth pattern changes were an increase in the complexity of the grinding surface, deposition of a bone-like substance called cement outside of the enamel, and an increase in the crown height of the teeth (hypsodonty). The net result of these modifications was a cheek dentition increasingly adapted for grinding by motion of the lower jaw from side to side against the upper jaw, for teeth that would remain free of deep pits as the tooth wore down, and for teeth that would endure years of grinding wear.

In competition with other herbivores in the environment, especially artiodactyls, equids became increasingly adapted to select and contend with the highest fiber, lowest protein diet in the grazing community by perfecting cecal (not ruminant) digestion in conjunction with increased rate of intake and passage (Janis 1976).

Other morphological changes were also occurring in the Miocene, although less rapidly than in the teeth. The skull was becoming more *Equus*-like, as was the brain. The eye was now set farther back in the head, and the muzzle was more elongated than before. The body and leg proportions differed between species; some were strong and stocky, others were slender and fleet in appearance. In adult *Merychippus*, the ulna had fused with the radius in the forelimb and was no longer movable as a separate unit. In the hindleg, the fibula had lost much of its shaft and was reduced to a spike of bone as seen in modern horses (see Figure 1.3). Such changes further limited rotation of the limb extremities. The limbs were specialized for locomotion with spring-like action, moving only in a fore-and-aft plane. The extremities did not retain maneuverability for holding or manipulating objects; yet fetlock flexibility was greater (Sondaar 1968). In the most advanced forms, the side toes were short and the primitive footpad of their ancestors had been lost. The weight was carried on the central toe which was tipped with a large convex hoof. *Merychippus* diversified into a number of descendant varieties; one major group migrated from North America to the Old World. Body size, side-toe length, and tooth pattern varied between species.

Apparently the only descendants of *Merychippus* where the side toes were finally lost was in the genus *Pliohippus*. Only internal vestiges of the side toes remained, these were long splint bones along each side of the cannon bone. Thus in the Pliocene epoch some equids, *Pliohippus*, were one toed. They were capable of swift, prolonged running. The teeth continued to increase in height and in complexity of the grinding surface. A unique feature not found in modern equids was the tendency for the skull of *Pliohippus* species and the descendant groups in South America to have deep pockets in the skull surface anterior and below the eye sockets. The function of such depressions is not clear. Possibly they housed some glandular material or sac-like extensions off of the nostrils.

Toward the end of the Pliocene, at least two million years ago, *Equus* arose apparently from the subgenus of *Pliohippus* called *Astrohippus*. This occurred in North America. There was some further lengthening of the cheek teeth which became straighter and somewhat more complicated in structural details; such trends continued as time progressed. *Equus* dispersed to the Old World via the Bering Land Bridge soon after the early forms appeared in North America and while still in the more primitive stage. The spread to South America over the Panama Bridge soon followed.

In the two million years since their first appearance, members of the genus *Equus* have migrated in many different directions and at different times. Each of the many species have had their distinct form and, no doubt, distinct habits. Throughout much of North and South America, Europe, Asia, and Africa fossil remains of *Equus* occur widespread and abundant in Pleistocene deposits. In both of the Americas, these equids survived the Ice Age and were still common

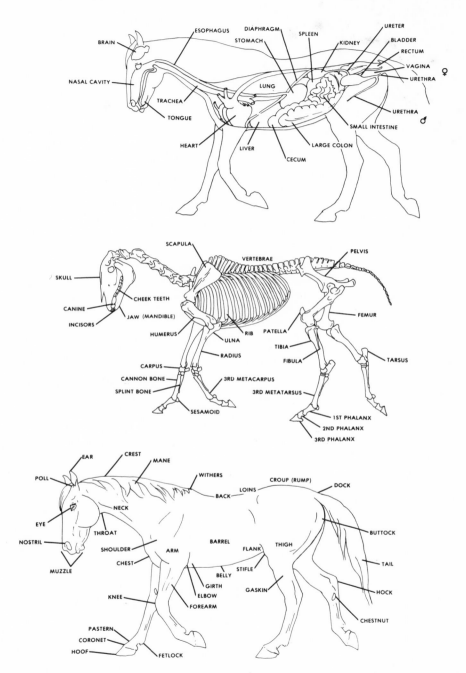

Figure 1.3: Internal and external morphological characteristics of the horse.

when the first Indians arrived, but then the herds on both American continents mysteriously and completely disappeared.

Equus species survived and diversified in Asia, Europe, and Africa. A definitive representation of the recent progression of horse evolution is difficult from only the fossil record; it is difficult to separate even the living species of *Equus* using only bones and teeth. To identify living species, we rely heavily on pelage characteristics, such as color and markings. Thus it is easy to understand the problems investigators have in precisely tracing the more recent lineage of our modern equids.

In late Pleistocene prior to the beginning of domestication, long-term isolation of equid populations undoubtedly occurred which led to what is now distinct species. The true (caballine) horses inhabited Eurasian lowlands north of the great mountain ranges. The hemiones occupied the arid zones of Asia from the Gobi to Syria and into northwest India. The true ass ranged primarily along the northern zone of Africa (Zeuner 1963). While each species continued to evolve characteristics independent of the others, they also differentiated into geographical races or subspecies which are now more or less distinct. These geographical races are apparent in the hemiones where four extant subspecies are recognized (see Table 1.1). The mountain zebra occurs as two subspecies; the Burchell's zebra, as three contemporary subspecies; and the ass, as two races in the wild condition. The surviving caballine horses are now reduced to two kinds—the domestic horse and the Przewalski horse.

Some authors have suggested that domestic horses were derived from more than one wild population. Their motive has been to explain differences in conformation of the animals depicted in ancient cave paintings, engravings and sculptures as well as differences noted among contemporary and ancient horses, such as in body size, temperament, and other characteristics. For example, Speed and Etherington (1952a, 1952b, 1953) Ebhardt (1954, 1957, 1962), and Skorkowski (1956, 1971) have furthered the concept of a multiple origin of the domestic horse from several discrete primitive types present in the Pleistocene. Chronological gaps, cytogenetic issues, and alternative explanations based on selective breeding are often slighted in such essays. Unequivocal supportive evidence is lacking.

Evidence that is available from Paleolithic times to the present suggests that horses of quite possibly different types were widely scattered along the arid and the fertile steppes, forests and tundra of Eurasia. These herds probably belonged to a single species (e.g., Nobis 1971) and could potentially interbreed yet were remaining reproductively isolated until influenced by human activities. To account for the apparent scattered and intermingled distribution often noted in these horse types, Zeuner (1963) suggested that these populations maybe were not strictly geographical subspecies per se, occupying different land masses, but may have been ecotypes, preferring different habitats. Thus each variety would tend to occupy its preferred habitat type (i.e., grassland, loess-steppe, tundra, or forest) wherever the herds existed across Eurasia.

Apart from geographical or ecological separation, social behavior and herd organization may also have caused discrete population characteristics, such as color, to develop and be maintained. For example, prolific harem stallions showing a preference for mares of one color could increase the frequency of genes with that characteristic in subsequent generations. Linkage and pleio-tropism could carry along additional genetic characteristics. Feist (1971) noted that feral horses he observed showed evidence of distinct color preferences. Some stallions had only buckskin mares in their social units; others had no buckskin mares but emphasized sorrel or bay. If descendants of those herds maintained similar preferences (e.g., through early experience) subsequent herds might emphasize one set of characteristics and other herds may empha-size different traits. Such differences would be similar to that observed in pale-ontological and archaeological records of Eurasia where varieties were not geo-graphically isolated in a distinct way or by good physiographical barriers. Thus social preferences, social attachment, and other behavioral traits of herd mem-bers could create a montage of different population characteristics throughout the distribution of the species.

Reproductive isolation by whatever means would account for some progres-sion of distinct morphological characteristics. The varieties surviving glacia-tion and present at the dawn of horse domestication were according to Zeuner (1963) and Heptner et al. (1966): (a) Przewalski horse, (b) tarpan, and (c) forest horse.

In reviewing and summarizing the domestication of the horse, Epstein (1971) concluded that pastoral tribes of the Mongolian steppes and plateau probably did not first domesticate the horse, but that domestication occurred in the early third millennium B.C. by a settled agricultural population in the western part of the grassland zone of the European Plain, such as the Tripolye culture in the valleys north of the Black Sea. Horse domestication and utiliza-tion seems to have been patterned after the well-established onager domestica-tion and use. The onager had been domesticated and used as a draft and occa-sional riding animal since possibly as early as the sixth millennium B.C. by Sumerians and neighboring cultures to the south. Wild horses did not occur in the more southern and Mediterranean regions where onager domestication first occurred. Horse herds did occur northward and were available to the Tri-polye and the Caucasus cultures. These horses are thought to have possessed coarse features, more characteristic of the Przewalski horse than any other variety (Epstein 1971, Brentjes 1972).

Heptner et al. (1966) suggested the zone between Prezewalski horses to the northwest and tarpans to the west was perhaps the Volga River. If so, the horses in the vicinity of the Tripolye settlements would have been tarpans. Furthermore, the range of the forest horse was north and westward of the Pinsk Marshes north of Kiev, quite accessible to the Tripolye settlements nearby. Thus the controversy as to which horse type was initially utilized in domestication remains complicated and unresolved.

Knowledge of horse domestication and use, once begun, spread rapidly through Asia and Europe, especially with the introduction in the early second millennium B.C. of the lighter weight horse-drawn war chariot with spoked wheels. Selective breeding was concurrent with the spread and diversification of horse utilization and included crossbreeding with ass and onager (Brentjes 1969, 1972).

Eventually wild varieties, with the exception of small remnant herds in in-accesible or barren environments, were absorbed into the domestic stock (Epstein

1971). Moreover, in time, wild and feral herds were systematically reduced or eliminated because of their depredation on agricultural crops and attempted covetry of domestic mares.

The traits emphasized in the domestic herds varied between cultures and as needs arose, such as mounts for heavily armored riders. Thus selective breeding can explain variations in size, facial appearances, color, temperament, and other characteristics that were noted in ancient as well as the more recent breeds.

The domestic horses of today may only partially resemble their wild ancestors in conformation and color; yet many traits are shared. Basic behavioral and physiological traits may have been little altered by domestication; domestic horses can still readily adapt to a wild existence. Feral herds show survival traits typical of species that have never been domesticated. Management practices may suppress certain behavioral tendencies, but the potential remains. In subsequent chapters, behavioral traits of horses under free-roaming conditions will be emphasized as the characteristics of the species.

2

Perception and Orientation

VISION

Horses are well known for their keen sensory perception. They are alert to changes in their environment and have long utilized their adept perception to facilitate survival. Undoubtedly the most important receptor system of horses is the visual system; nevertheless for this species, this system is yet to be fully investigated and is often misrepresented.

To begin to understand this sensory system, it is necessary to put equine vision in a proper ecological, morphological, and physiological context. Thus first consider the horse's need for vision. Horses, as were their recent ancestors, are basically open range animals with little threat from aerial predators. Their predators have been primarily ground dwelling forms, as are their social companions and their nutritional sources. Thus, it should not surprise us to find the equine visual system is tuned not only to a wide panorama of the horizon but also toward the front of the animal where it must place its feet, obtain nutrition, and avoid ambush as it moves. Its visual realm is not skyward but groundward. In their natural habitat, it is beneficial for horses to see in bright light as well as in the nocturnal period.

The eyes of horses are in a lateral position relatively far back on the skull. Each eye is rotated and moved synchronously with the other by the interaction of seven muscles attached to the eyeball. In addition, the eyes can be elevated, turned, and tilted by supplementary movements of the head and neck. At rest, the optic axis of each eye diverges about 40° from the anterior midline (longitudinal axis of the body) and about 20° below the horizontal (Hughes 1977).

The morphology of the equine eye is unusual not only in size but also in shape. The horse has one of the largest eyes of any living animal (Figure 2.1a). The retina is asymmetrical with a tendency for the retina to be closer to the lens especially, but not exclusively, below the optic axis (Nicolas 1930, Sivak and Allen 1975). In 1818, Soemmerring first illustrated this phenomenon and noted the distance between the cornea and the retina of horses (38mm) surpassed most other animals. Besides this large internal space, he observed the circumference of the horse retina was even greater than in the eye of the far larger bowhead whale (Andersen and Munk 1971).

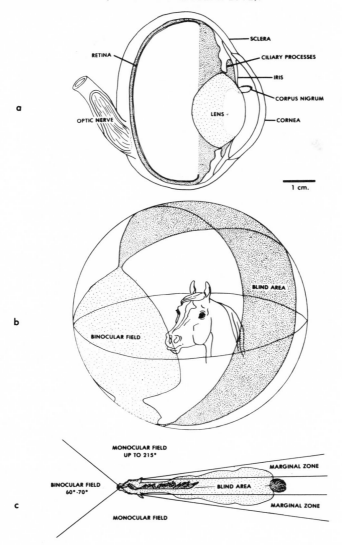

Figure 2.1: Vision in horses: (a) the asymmetrical eye (after Soemmerring 1818), (b) ophthalmoscopically defined ocular field (based on Pisa 1936 and Hughes 1977), (c) panoramic visual field (adapted from Waring et al. 1975).

The expansive retina of the horse allows for an extreme range of peripheral vision. Each eye has a horizontal visual field of up to 215° (average 190°-195°).

An overlap of the visual field of each eye occurs, giving the horse a 60°-70° binocular field of view anteriorly (Duke-Elder 1958). This binocular field of view is extended downward along the midsagittal plane (Figure 2.1b), enabling the horse to view the ground in front with both eyes (Figure 2.2). The retinal field of view of each eye in the vertical plane is 178° (Hughes 1977). In the posterior direction, the visual field almost parallels the body axis leaving a narrow blind zone behind the animal (Figure 2.1c). Of course, a slight turn of the head enables the horse to scan even this area behind its body. In strong light the vertical diameter of the pupil narrows, accentuated by the corpus nigrum, forming an oblong horizontal opening which reduces light yet maintains the visual field in the horizontal plane.

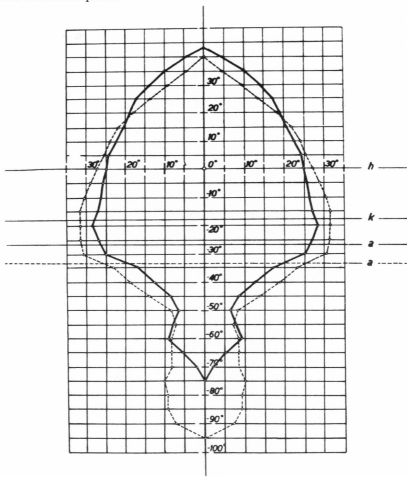

Figure 2.2: Projection of the binocular visual field of an adult horse (solid line) and a foal (broken line). Maximum width of the binocular field coincides with plane (k) formed by the corners of each eye. Plane of optic axis = a; horizontal plane = h. (From Pisa 1939)

Histological examination shows the retina to be a complex consisting of numerous microscopic layers. Among the neural elements are rod and cone receptor cells. Although the equine retina lacks a depression called a fovea, it does have two regions of acute perception. The region corresponding to the sensitive macula lutea in man is called the area centralis (or area retinae) and is located about 15 mm dorsal and slightly lateral to the optic papilla, where the optic nerve merges with the retina. This optically acute area is 2 to 5 mm in diameter (Prince et al. 1960, Prince 1970). The second area of increased ganglion-cell density is associated with the area centralis and is a band-like area called the visual streak which extends horizontally across the retina. It lies about 8 mm dorsolateral to the optic papilla and medial to the area centralis (Prince et al. 1960, Hughes 1977). In both the visual streak and the area centralis, there is a higher proportion of cones to rods, although a duplex retina is maintained. The area centralis functions especially in forward binocular vision, and the visual streak appears to broaden horizontally the acute field of view.

Although daylight vision is most acute, nocturnal vision in horses is superior to that of man. Hughes (1977) has calculated (based on maximum pupil diameter) that the horse, owl, dog, and gray squirrel have similar light collecting power in spite of the difference in the size of their eyes. They rank below the cat, rabbit, rat, and bat. Similar to many nocturnal animals, the horse eye has a developed tapetum lucidum, a fibro-elastic tissue which reflects light back through the retina and causes eyeshine when the eyes are illuminated at night.

The type of accommodation or focusing mechanism occurring in horses has been the subject of controversy and needs further study. Based on the asymmetrical retina (usually considered skewed with greater distances occurring in the dorsal direction), some authors conclude a static accommodation system is operating where nearby images and distant images simultaneously focus on different parts of the so called "ramp retina." They suggest no dynamic or lens-adjusted accommodation occurs (e.g., Walls 1942, Duke-Elder 1958).

The existence of the area centralis and associated visual streak (sites of acute vision) seems to contradict the concept of a functional ramp retina; yet such sensitive sites are consistent with dynamic accommodation. Moreover, Sivak and Allen (1975) could find no indication of a ramp retina that could serve accommodation and, in fact, observed some dynamic accommodation ability in living horses. Without excluding either concept, Prince et al. (1960) suggested that a small degree of ciliary accommodation (dynamic) could exist together with the ramp retina in the horse.

The equine lens is elastic–a necessity in ciliary accommodation. Thus it is proposed by Sisson and Grossman (1953) that to accommodate for near objects, the ciliary muscle contracts and pulls the ciliary processes and associated ciliary ring forward, releasing tension on the lens and thus slackening the ciliary zone, allowing the lens to become more convex (see Figure 2.1a). Motion of the midpoint of the lens forward (axial translation) plays some role in accommodation of some species; however, this mechanism, though feasible, has yet to be demonstrated in horses (cf. Hughes 1977). In old age, the lens tends to become less elastic and may lose its transparency.

As a result of the optical and morphological properties of the equine eye, motion along the edge of the field of vision may be accentuated (Simpson 1951,

Knill et al. 1977). Undoubtedly, some ganglion cells of the retina are specialized to help detect peripheral motion, such as may be made by a predator. A horse is often startled and overreacts to the apparent sudden motion occurring on the ground at the margin of its visual field while standing or as motionless objects momentarily appear in and out of the visual field while the horse itself is moving. These responses were likely of survival benefit to wild ancestors who were vulnerable to predators. Sudden flight was the best defense.

The debate continues over the existence of color vision in animals such as the horse. In general, investigators are finding color perception is more widespread than formerly realized. The horse retina does have both rod and cone receptors, and color detection is considered to involve especially the cones. Grzimek (1952) investigated color vision in two mares, four and six years of age. He concluded his subjects could see color and not merely different shades of gray. In a series of discrimination trials contrasted with 27 shades of gray, the yellow test colors were identified most easily, green colors were second, then the blues, and least the red colors. Light red was selected more easily than more absolute red choices; yet, saturated blue choices were correctly chosen more easily than lighter blue. At 3.3 m, a 0.5 cm perpendicular yellow line was reliably detected (a visual angle of as little as 3'15"); whereas at the same distance, a blue streak of a minimum of 2 cm was reliably detected (angle of 20'41"). Thus the acuteness of vision of the horse appears to be slightly less than that of humans; our eyes also are limited in blue visual acuity.

Besides color detection, horses show good visual pattern discrimination. They can learn to recognize correct choices in twenty or more two-choice discrimination sets, such as triangles verses dots of the same size (Dixon 1966). In the recognition of human beings, horses rely on facial characteristics as well as clothing (Grzimek 1944b). Anecdotal literature reports that some Arabian horses have been known to visually identify their master from similarly dressed men at a distance of 0.4 km (0.25 mile) or more.

Consistent with their discrimination abilities and being social animals, horses respond to horse-like objects differently than they do to other test objects. Grzimek (1943a) found two- and three-dimensional horse imitations were approached and investigated like conspecifics, for example, at the nose and flanks; but incomplete drawings and dog pictures were not. Vision is used for individual recognition between horses along with odors and vocal characteristics (cf. Wolski et al. 1980).

Additional indications of the visual acuity of horses are the fascinating stories of such horses as Kluge Hans (Pfungst 1907), Lady (Rhine and Rhine 1929a,b), Muhamed, and Mahomet (Christopher 1970). These horses amazed observers by answering mathematical, spelling, and other questions with head movements and leg gestures. Yet in each case, it was eventually discovered that the horses could only perform accurately if someone was present who knew the answer and signaled the solution by a slight gesture to the keenly observant horse.

HEARING

Horses have been reported to respond to low frequency geophysical vibrations (possibly P waves) preceding the shaking of earthquakes and to high frequency sounds above the human perception range. Moments before the ground starts to shake, horses often show nervousness and vocalize (e.g., Lawson 1908, Penick 1976). Ödberg (1978) in a test of equine hearing, observed distinct ear reactions (Pryer reflexes) to pure tones at freqeuncies up to 25 kHz (25,000 cps). At the highest frequencies, older horses (age 15-18) showed less response than subjects of 5-9 years of age. It appears that horses are able to detect a broader range of sound vibrations than can humans; nevertheless, the bulk of the sound energy perceived by horses is most likely within the frequency and amplitude range audible to human ears.

It becomes readily apparent to an observer that horses rotate their ears (pinnae) in response to directional sounds. The independently movable pinnae enable acoustical orientation toward sound sources without the necessity of changes in head or body position. A complex of muscles innervated by branches of the facial as well as first and second cervical nerves induces the action of the ears. When the ears are vertical and drawn forward, the opening is directed forward. The opening also can be rotated to focus to the side or posteriorly; whereas, when the ears are fully laid back, the opening is toward the ground and closed by compression.

TOUCH, PRESSURE AND THERMORECEPTION

Tactile or touch perception occurs over most of the horse's body, with especially sensitive areas around the head. As handlers readily discover, horses avoid tactile stimulation in and around their ears. Innervation of hair follicles is widespread and often involved in tactile responses. Specialized, stiff tactile hairs with sensory innervation at their base project beyond the remaining hair coat; these hairs are especially prevalent around the lips, nose, and eyes (Talukdar et al. 1972).

Sensory end organs occur in different forms within the skin. For example, in the dexterous upper lip of horses, three groups of sensory nerve endings are found: (1) endings with an inner core (lamellated and encapsulated), (2) endings with auxillary cells (non-lamellated but sometimes encapsulated), and (3) free nerve endings (Talukdar et al. 1970). Capsulated endings seem to be limited to the dermis. Lamellated endings are oval and are covered by a thin capsule composed of one layer of cells. Within these capsules, a single lamella of squamous-like cells surround the nerve fiber in the center. The non-lamellated yet encapsulated endings are largest in the deeper layers of the dermis; disc and spray-like endings occur. Free nerve endings occur in the superficial dermis as well as into or just below the stratum granulosum of the epidermis. Such sensory end organs are thought to be associated with touch, pressure, and thermoreception.

SMELL AND TASTE

Chemoreception in horses involves at least three receptor systems: (1) the olfactory nerve endings of the nasal cavity, (2) the vomeronasal organ, and (3) the taste buds. The olfactory nerve endings commonly associated with smell are located toward the posterior end of the elongated nasal cavity, specifically on the lateral masses of the ethmoturbinates, the adjacent part of the dorsal turbinates, and the septum nasi. The elongated olfactory cells are situated between supporting cells in a yellow-brown, non-ciliated epithelium. A tuft of fine, hair-like filaments extends from the olfactory cells into the nasal cavity. The other end of the cells form non-medullated nerve fibers leading to the olfactory bulb (Sisson and Grossman 1953). Literally millions of the olfactory receptor cells may be present.

The paired vomeronasal organ lies beneath the floor of the nasal cavity along each side of the anterior lower border of the nasal septum. The two parts of the organ extend posteriorly as blind-ended cartilagenous tubes about 12 cm long. Both tubes are lined with mucous membrane and contain sensory fibers of the olfactory nerve. At the anterior end, the tubes communicate with the nasal cavity by a slit-like orifice in common with the incisive or naso-palatine duct (Sisson and Grossman 1953). Since in *Equus* species the duct communicates only with the nasal cavity, odorous chemical substances enter the vomeronasal organ via the nasal cavity.

Estes (1972) suggested that use of the vomeronasal organ is facilitated by an animal filling its nasal cavity with odor ladened air (such as in urine testing), closing the external nares through the flehmen response (see Figure 2.3b), and allowing the moist air to flow into the vomeronasal ducts upon elevation of the head above horizontal. Gravity, vasomotor movements, or perhaps vasodilation of the organ assists in the induction of odorous chemicals. Liquid-borne compounds of low volatility are most likely involved. Stallions seem to flehmen more than other sex/age classes. Functional significance pertains to social interactions. Estes (1972) concluded the function of the organ is primarily for the detection of estrus and the release, control, and coordination of sexual activity by measuring levels of sex hormones or their breakdown products in the discharges of companions.

Olfactory cues supplement other sensory cues and, therefore, can affect behavior. Wierzbowski (1959) found that when urine from estrous mares was sprinkled on a semen collecting dummy, young stallions showed an increased sexual response. Both mares and their foals recognize each other in part by odor. Disruption of both olfactory and visual cues can greatly impede their process of individual recognition (Wolski et al. 1980).

Chemoreception by taste involves the microscopic taste buds innervated by fibers of the glossopharyngeal nerve and the lingual branch of the trigeminal. The taste buds occur especially on the foliate, fungiform, and vallate papillae (Figure 2.3c) of the tongue as well as on the free edge and anterior pillars of the soft palate and the oral surface of the epiglottis (Sisson and Grossman 1953). The taste buds are barrel-like masses of taste cells embedded in the epithelium. Each bud has a minute opening called the gustatory pore through which small filaments, the microvilli of the taste cells, project.

The taste sensations perceived by the horse are presumed to be gradations of salt, sour, sweet, and bitter. Yet taste is not easy to analyze in animals and varies between species as well as between individuals. For example, the horse

does not differentiate between pure water and an aqueous solution of sucraocta-
acetate at concentrations which would be offensively bitter to us (Kare 1971).

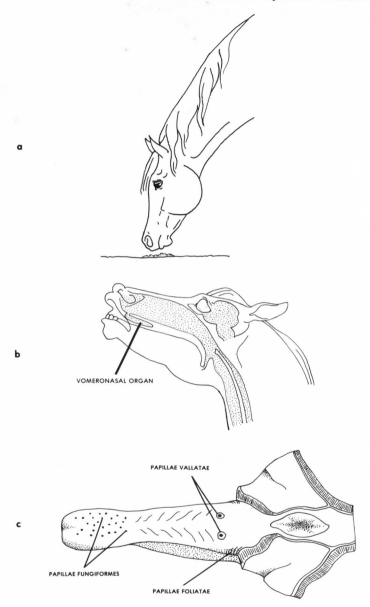

Figure 2.3: Chemoreception in horses: (a) typical olfactory investigation, (b)
possible involvement of vomeronasal organ during flehmen, (c) papillae where
taste buds are located on the tongue.

Quinine solutions are rejected by horses once the concentration reaches 20 mg per 100 ml (Randall et al. 1978). The latter study, using foals as test subjects, found sucrose solutions were preferred to tap water at concentrations ranging from 1.25 to 10 g/100 ml; below as well as above this range indifference was shown. The foals were indifferent to salt (NaCl) solutions until concentrations reached 0.63 g/100 ml; rejection then became typical as salt concentration increased. Sour perception was tested using acetic acid solutions; these solutions were rejected once the concentration reached 0.16 ml/100 ml and pH 2.9. Compared to other domestic species, foals respond most like sheep to sweet, salty, sour, and bitter solutions.

Horses choose and sort their foods using chemoreception and possibly tactile and visual characteristics. In this manner, poisonous plants are often avoided. The ability to choose appears to improve with maturity, but selective feeding varies depending on management practices, previous feeding opportunities, and factors such as season, hunger, condition of plant, time of day, and genetic background of the animal (Marinier 1980).

PROPRIOCEPTION AND EQUILIBRIUM

As in other mammals, horses have muscle and tendon receptors that provide the central nervous system information on the extent of stretch of the muscles and tendons. Such proprioceptive sensations provide the horse with information on the position of the various parts of its body without the need to monitor those body parts visually.

Equilibrium receptors are located in the inner ear embedded in bone along the temporal region on each side of the skull. These receptors consist of three fluid-filled loops (the semicircular canals) and two adjacent sac-like chambers, the utricle and saccule (Figure 2.4). The semicircular canals, arranged perpendicular to each other, have at one end a spherical expansion (ampulla) that is lined with sensory hairs. If the animal or its substrate moves or changes direction or speed, the fluid in one or more ampulla and canal moves past the sensory hairs causing a neural impulse to be transmitted up the vestibular nerve.

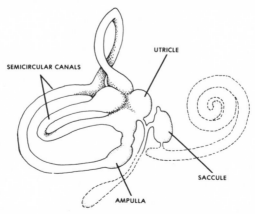

Figure 2.4: Equilibrium components of the inner ear.

Thus the individual is made aware of motion as well as changes in direction and speed.

Sensory areas in the utricle and saccule appear as small whitish thickenings composed of sensory cells with hair-like processes surrounded by support cells. Adhering to the surface of these receptors are fine crystals of lime salts. As the head tilts, the crystals shift due to gravity and stimulate the sensory cells and, thus, the vestibular nerve. The saccule also has a branch of the cochlear nerve which suggests low frequency vibrations may be detected here as in some other vertebrate animals.

PAIN

Observations suggest that pain in horses ranges from minor discomfort to extreme pain. Changes in posture and facial expressions, vocalizations (e.g., groans), loss of appetite, sweating, muscle tremors as well as increase in pulse and respiration rate are among the indicators of pain (Müller 1942, Seiferle 1960, Walser 1965, Fraser 1969). Pain is exhibited widely throughout the body and seems to not need any special organs for its reception; various physical and chemical stimulation direct to end fibers of sensory nerves can cause the sensation. Interoceptors, sensory fibers within the viscera, are involved in the discomfort evident during such ailments as colic.

Under stressful situations, individuals react differently to painful stimuli than they do in other situations. Selective perception or stimulus filtering seems to be involved. For example, responses to painful stimuli are diminished in foals during parturition until the pelvis has passed the vagina (Rossdale 1967a). Certain restraint techniques, such as using a twitch, appear to function because the strong stimulus being applied causes reaction to other stimuli to be temporarily reduced.

ORIENTATION AND HOMING

Orientation in horses is an interplay of three main types—stabilization of posture and movement, object orientation, and orientation in the environmental context. The eyes, proprioceptors, and semicircular canal complex enables stabilization of posture and movement independent of locality. When a horse moves toward or away from an object, any one or a combination of receptor systems may be involved, such as the visual, auditory, olfactory, or tactile system. It is common that a horse initially uses vision in its approach, then touch and smell, and finally taste when selecting food items.

Within their environmental context, horses normally remain well oriented. The effects of gravity on the sensory structures of the utricle and saccule of the inner ear as well as on proprioceptors provide the animal with cues to maintain its physical balance against the pull of gravity. Visual cues provide further information about the location of landmarks, the horizon, the sun, and the stars. Tactile and kinesthetic cues inform the horse of the type of substrate and how far it has traveled. On windy days, horses tend to orient parallel to air currents while feeding and resting.

Thus, a variety of sensory input and motor responses are involved in orientation. Whenever flight is necessary, or food, water, or shelter required, the appropriate direction is taken. Trails are often established and utilized along frequented routes. Long distance travel, such as to a water source 10 km away, appears to involve memory and landmarks.

Horses tend to follow the route of other horses. Vision undoubtedly is used to stay in the trail of a preceding horse. Yet Janzen (1978) observed an instance where smell also seemed to be used by a horse to verify the trail made about an hour earlier by a companion horse that had already traversed the coastal beach. The trailing horse kept its nose about 1 cm from the sand while searching for and verifying the intermittent trail at the water's edge. Each time tracks became evident, the horse raised its head to a normal height only after smelling the trail for at least 100 meters.

The tendency for horses to return home when given free rein has long fascinated horsemen. Homing seems to result from a combination of the desire to stay in a specific home range with known resources and a desire for social contact. Grzimek (1943b) and Williams (1957) attempted to investigate homing by allowing individual horses free choice of travel several kilometers from their home stable. The results were not supportive of a well-developed directional sense; nevertheless, the horses made attempts to locate surroundings familiar and congenial to them. Williams noted that there was a tendency for orientation into the wind regardless of the direction of home. Smell is important in social behavior and probably assists in homing. The ability of horses to utilize existing cues, such as faint odors, in addition to memory as well as trial and error, appears responsible for their homing capacity. A more elaborate navigation system in horses has not been demonstrated.

3

Motor Patterns

REFLEXES

Among the most basic motor patterns of horses are those associated with reflex arcs. They involve relatively few sensory-motor units, keeping the neural as well as muscular involvement simple. Since reflexes are stereotyped and involuntary responses to a given stimulus, they are useful to veterinarians to determine not only stages of anesthesia but also soundness of the neurological system (Rooney 1971, Catcott and Smithcors 1972). For example, a slight tap of the side of the neck just posterior to the ear will cause the ear on that side to turn forward (cervico-auricular reflex) provided the sensory components of the most anterior cervical tracts and the pathway through to the facial nerve and auricular muscles is functioning properly (Rooney 1973).

Associated with the head are a variety of reflexes of the eyes, ears, nose, and mouth (Table 3.1). Tapping the bone just below the eye, for example, causes the palpebral or eyelid reflex shown as a blink. An object visually perceived nearing the eye as well as corneal stimulation causes additional blink reflexes. Material in the eye induces the lacrymal reflex. Sudden bright light causes pupil constriction called the pupillary light reflex. Tonic eye reflexes keep the eye looking in the original direction when the head is moved. Distinctive sounds cause ear twitching characteristic of the Pryer reflex. The head shake reflex occurs with tactile stimulation of the hairs of the ears. Tactile stimulation around the mouth of a newborn foal initiates the sucking reflex. As solid foods are eaten by horses, the mastication reflexes modify chewing activity to protect the tongue and other tissues from harm. The salivary reflex occurs when material enters the mouth. Stimulation of the nasal mucosa causes the sneeze reflex, whereas stimulation of laryngeal mucosa induces the cough reflex.

Some reflexes are associated with posture and body orientation (cf. Rooney 1971). With tilting the head up or down and not altering the neck position, the vestibular reflexes occur; upward head extension tends to flex forelimbs and extend hindlimbs, whereas ventroflexion of the head induces hindleg flexion and foreleg extension. If the head is kept in a normal position and the neck only is moved, then the tonic neck reflexes occur; dorsiflexion of the neck tends to flex the hindlimbs and extend the forelimbs, whereas ventroflexion of the neck causes forelegs to flex and hindlegs to extend. Pressure on the soles of the feet causes leg extension (extensor thrust reflex), thus a horse stands without conscious thought. When pressure is applied to the side of a horse, the near legs tend to flex and the opposite legs extend (sway reflexes involving crossed extensor response). Pressure on the croup induces flexion or tucking at the lumbosacral joint. Pressure at the lumbosacral junction causes upward tilting of the pelvis and hindleg extension; pressure near the thoracolumbar junction promotes dorsiflexion of the back (vertebra prominens reflexes). Labyrinthine reflexes reacting to gravity are involved during righting responses, such as when a horse lying on its side raises and levels the head with the neck twisted to achieve sternal recumbency. Supporting and placing reflexes also assist posture and coordinated motor activity, such as the segmented static reflexes where as one leg leaves the ground the other legs extend in response.

Table 3.1: Some Reflexes of the Horse

Eye, Ear, Nose and Mouth Reflexes–
    Palpebral reflex
    Corneal reflex
    Lacrymal reflex
    Visual blink reflex
    Pupillary light reflex
    Tonic eye reflexes
    Cervico-auricular reflex
    Pryer reflex
    Head shake reflex
    Sucking reflex
    Mastication reflexes
    Salivary reflex
    Sneeze reflex
    Cough reflex

Postural Reflexes–
    Vestibular reflexes
    Tonic neck reflexes
    Sway reflexes
    Vertebra prominens reflexes
    Labyrinthine reflexes
    Segmental static reflexes

Miscellaneous Reflexes–
    Panniculus reflex
    Abdominal cutaneous muscle reflex
    Perineal reflex
    Local cervical reflex
    Withdrawal reflex
    Kicking reflex
    Bucking reflex
    Thrusting reflex
    Ejaculatory reflex
    Spinal visceral reflexes

Additional reflexes occur throughout the body for various other biological functions. For example, the panniculus reflex causes twitching of cutaneous musculature when the skin is pricked or stimulated by biting insects. Tactile stimulation of the hairs along the costal (rib) arch induces cutaneous muscle contraction, especially of the flank (abdominal cutaneous muscle reflex). Tactile stimulation of the tissues near the anus causes the perineal reflex; the anal sphincter contracts and the tail is clamped down (except in estrous females or those nearing parturition). Tapping the side of the neck between cervical vertebrae 3 and 5 causes local muscular contraction (local cervical reflex). A noxious stimulus applied to the distal portion of a limb causes the withdrawal reflex. Moving a hand along the hindleg of a foal tends to cause the kicking reflex, whereas moderate pressure on the kidney region of the back induces bucking in very young foals. Grzimek (1949a) found the latter response disappeared on the eighth day of age in the foal he studied. As stallions develop sexually, the thrusting reflex of the pelvis eventually accompanies mounting, and associated with high sexual excitation is the ejaculatory reflex. Spinal visceral reflexes control urination and defecation.

LOCOMOTOR ACTIVITY

Based on fossil evidence, postures and movement patterns of equids have changed as a result of evolutionary changes in body and limb morphology (cf. Sondaar 1969). Locomotor characteristics have changed concurrently with anatomical and physiological changes. With body size increases, it became necessary for alterations to occur in the proportions of the running apparatus in order to retain swift locomotion. Thus, the movement patterns of the domestic horse are the result of millions of years of selective processes. The reduction of toes until the single central digit supported body weight led to corresponding adaptations in leg structure. Besides lengthening of the limb bones and changes to prevent lateral movement of joints, the development of the so-called spring ligaments was significant. With these elastic ligaments and with maximum flexibility of the fetlock in the anterior-posterior direction, a spring-like mechanism resembling the effect of a pogo stick was created. Within limits, the harder the impact on this apparatus the higher the bounce. This type of foot was obviously very effective on firm soil, extended the endurance of the animal, and permitted a size increase yet maintained speed. Locomotion continues to be fundamental to every horse.

Locomotor activity in healthy foals begins within minutes following birth and continues to serve biological needs throughout the life span. In a horse's world, little can be accomplished without moving about. For example, Feist (1971) found the feral horses he observed along the Wyoming-Montana border sometimes needed to travel each day as much as 16 km from their feeding site to reach a water hole for a drink. A newborn foal must stand and move about to search for its first meal, just as an older horse must move about to feed and obtain water. Soon after birth, a foal can travel swiftly with its mother for short distances. In their prime, some grown horses can attain a speed of more than 65 kph (40 mph) for nearly a kilometer. A distance of 32 km (20 miles) can be covered in one hour by many horses (Hildebrand 1959); the pony express horses in the American west during 1860-1861 demonstrated such endurance.

Normal locomotor activity of horses can be inhibited by physical, chemical, and psychological restraint or disrupted physiologically as a result, for example, of trauma, infections, or toxicity. An active horse subsequently restricted in forward locomotion often exhibits pawing, seemingly as a displacement act (Ödberg 1973). Littlejohn (1970) noted that horses recovering from general anesthesia spent fractionally more time walking but walked much more slowly during the first 30 minutes of standing than when normal.

The area of the cerebral cortex where somatic motor activity can be elicited occupies nearly the entire rostral half of the dorsal surface of the cerebral hemispheres. Electrical stimulation of this part of the brain has shown there are four distinct motor regions. That is, stimulation with electrodes from anterior to posterior causes (1) contralateral upper and lower lip movement, (2) contralateral nostril dilation, (3) contralateral shoulder and neck movement, and (4) contralateral limb movement (Breazile et al. 1966). The last area is especially important in locomotion; such motion is coordinated through the cerebellum. Horses show a slight individual, but not species specific, tendency toward either right or left handedness (Grzimek 1949b).

Not only do the legs have a role in locomotion, the neck, spine, and associated muscles play a part as well. Various parts of the body together with the vestibular apparatus of the ear also have a role in postural reflexes. Relative to many smaller mammals, the equid spine arches little during galloping strides; the withers to ground distance remains relatively constant, the croup to ground distance varies only slightly, and the chest to buttock length changes only moderately during a stride (Hildebrand 1959). Nevertheless, during a vigorous gallop (13 m/sec), the angular displacement of the neck can be 28° with this variation occurring systematically during the stride (Figure 3.1). The downswing of the neck begins during the suspension or flight phase of the stride and continues in a nearly linear manner as the first three feet contact the ground.

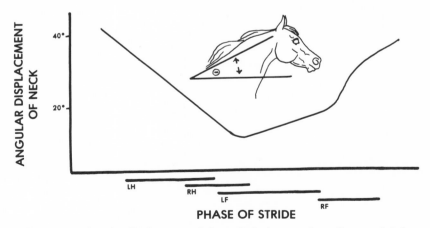

Figure 3.1: Angular displacement of the neck during a racing gallop correlated with the support periods of the individual legs—left hind (LH), right hind (RH), left fore (LF), and right fore (RF). (After Rooney 1978)

As the hindlegs leave the ground, the neck begins the upswing, reaching a maximum as or shortly after the lead foreleg leaves the ground. Rooney (1978) has postulated that if the muscles forming a mechanically continuous system from the cervical to the thoracic vertebral column (Figure 3.2) were to hold in isometric contraction, then as the neck moves down, the body in effect would be pulled forward contributing to linear forward motion of the horse.

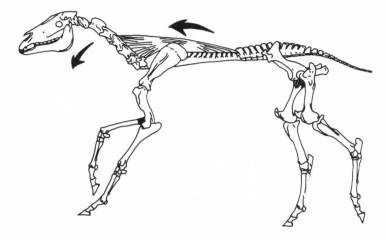

Figure 3.2: Musculature involved in lifting the body forward as the neck moves downward. (After Rooney 1978)

Gaits

The specific gaits of horses reflect not only anatomical characteristics but also a balance between energy expenditure and environmental context. The walk permits locomotion with a minimum energy output. When more speed is necessary, the trot or a slow gallop is used. And finally for bursts of extreme speed a vigorous gallop occurs (cf. Tricker and Tricker 1967). The natural gait at any speed entails the smallest possible energy expenditure (Hoyt and Taylor 1981). Heglund et al. (1974) found that stride frequency and stride length both increased with increasing speed; however, within a gallop, speed was increased primarily by increasing stride, whereas frequency remained nearly constant. Because of the nearly constant stride frequency in the gallop, Heglund and his co-workers concluded the transition from trot to gallop occurs at the maximum sustained stride frequency of the animal.

The gaits of walk, trot, and gallop are natural to all horses. Gaits such as the slow gait, running walk, and rack are considered acquired since the horse industry has selected for and trained certain horses to perform such movements. The pace can be either natural or acquired. As might be expected, gradations occur as a horse moves from one gait to another or changes speed. Terminology pertaining to gaits varies considerably from one breed association to the next and from one geographical region to another. No attempt in this book will be made to cover them all.

Walk: The *walk* typical of horses is what Magne de la Croix (1936) called a diagonal walk. All the limbs move sequentially one after the other as follows: left fore, right hind, right fore, left hind. Since each hoof hits the ground individually (Figure 3.3a), a walking stride consists of four beats. There is an alternation of between two and three feet supporting the body weight during this gait. In an ordinary stride, the hindfoot more or less covers the print made by the forehoof on the same side; a tired horse will usually place the hindfoot short of the impression made by the forefoot. Saddle horses tend to have a stride of about 5.75 m and average approximately 6.5 kph at the walk (Grogan 1951). From 0.6 to 1.0 stride per second is common (Hildebrand 1965).

A variation of the walk occasionally seen in horses is the lateral walk (Figure 3.3b) which often progresses into the pace. In this variation the first foreleg to move is followed by the hind on the same side, then the other foreleg moves and finally the diagonal hind.

Trot: The *trot* is a two-beat gait in which the two diagonal feet work as a pair and are either in the air or on the ground at the same time (Figure 3.3c). The footfall pattern is: (a) left fore, right hind, (b) right fore, left hind. The trot provides the animal with greater balance than does the pace. After each diagonal pair leaves the ground there is normally a brief moment where the animal is not supported by any legs until the other pair makes contact again with the substrate. The animated form of the trot, such as shown by a displaying stallion, is often called *prancing.* The speed of the trot commonly is in the range of 10-14 kph (6-9 mph); racing trotters, however, often average 50 kph (30 mph). During the trot, the legs bend more than in the walk.

Pace: The *pace* or amble is another two-beat gait where legs on the same side act in unison (Figure 3.3d). As in the trot, there are two periods during each stride where the body is suspended or in flight. The footfall pattern is : (a) left fore, left hind, (b) right fore, right hind. The pace is not a natural gait for most horses. When it does occur or is taught, its speed is similar to that of the trot.

Gallop and Canter: The *gallop* (Figure 3.3f) and its more restrained version, the *canter* (Figure 3.3e), are basically four- and three-beat gaits, respectively. A diagonal or transverse gallop best describes the footfall pattern typical of horses versus the lateral or rotary gallop typical of, for example, rabbits or the cheetah. As a cantering or galloping horse proceeds to contact the ground following the flight phase, it uses one of the following mirror image patterns:

I. a) left hind	II. a) right hind
b) right hind, left fore	b) left hind, right fore
c) right fore*	c) left fore*
*lead foreleg	

In the typical canter, the second and third legs contact the ground simultaneously; in a full gallop, the hindlimb of that pair (b) contacts the ground first, giving a four-beat rhythm instead of a three-beat. The body pivots over the lead foreleg, and it is the last leg lifted before all legs are again off the ground. As a horse leans or turns to the right, it normally leads with its right foreleg, and if motion is leftward so also is the lead on the left. Lead changes

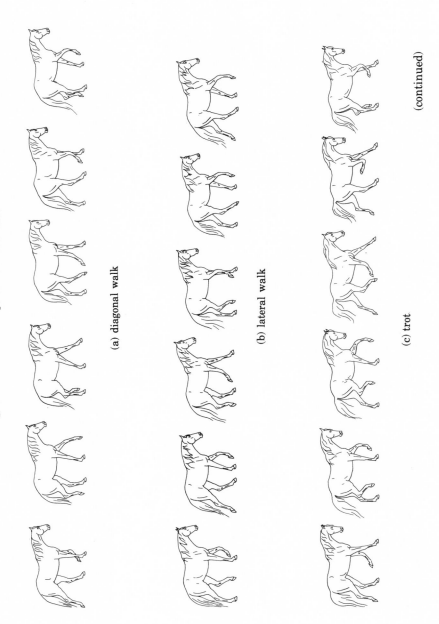

Figure 3.3: Locomotor patterns in the horse

(a) diagonal walk

(b) lateral walk

(c) trot

(continued)

(d) pace

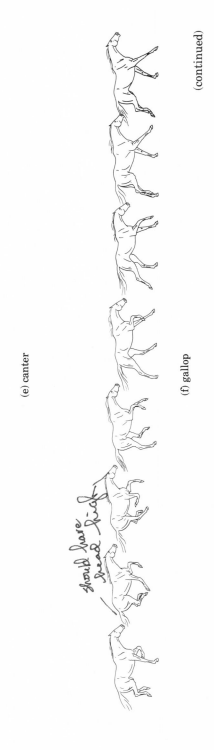

(e) canter

(f) gallop

(continued)

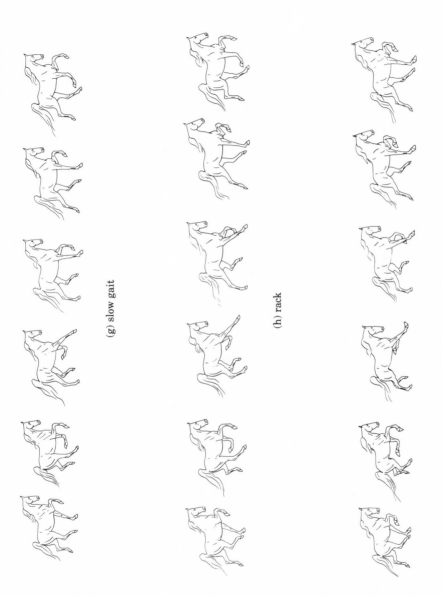

(g) slow gait

(h) rack

(i) running walk

can occur by switching the footfall pattern of the forelegs subsequent to the hind contacting the ground or more commonly during the suspension phase of the stride by placing the other hindleg down first. Riders often school their horses to change leads upon command.

A slowed gallop becomes a canter. As the canter is slowed or animated considerably, the three-beat characteristic may shift to a four-beat pattern as the second and third legs begin to contact the ground separately rather than together. The speed of the canter is approximately 16-19 kph (10-12 mph); a more natural gallop would be about 26-29 kph (16-18 mph), with a maximum racing gallop of 64-69 kph (40-43 mph). In the gallop, a horse covers usually 5.8 to 7.6 m per stride and at 56 kph (35 mph) completes about 2.3 strides per second (Hildebrand 1959, 1977).

Slow Gait: The so-called *slow gait* of five-gaited show horses is one of the acquired gaits (Figure 3.3g). It is generally a very animated, slow, broken pace (sometimes called a stepping pace). Although the legs of the same side leave the ground together, the hind returns before the high-stepping foreleg. One to three legs support the body at any one time.

Rack: The *rack* is another of the gaits of five-gaited show horses (Figure 3.3h). It follows the leg pattern of the walk but is faster and more animated. High foreleg action occurs. The rack is particularly fatiguing for the horse. Hildebrand (1965) noted that a given horse may complete 1.6 to 1.8 strides per second in a slow gait and 2.0 to 2.1, at the rack.

Running Walk: The *running walk* is the acquired gait distinctive of the Tennessee Walking Horse. It is the fastest of the four-beat show gaits, exceeding 32 kph (20 mph). The gait has a smooth gliding motion with forelegs extending greatly (Figure 3.3i). High animated leg lift is not typical. Enormous steps are taken in an accelerated walk footfall pattern. The head and neck nod up and down as the forelegs are advanced. Hildebrand (1965) found 1.5 to 2.2 strides per second were completed in the running walk.

OTHER MOTOR PATTERNS

Two Tracking: Horses show a great variety of motor patterns other than those already mentioned (cf. Table 3.2). For example, during a walk or slow trot, a horse can shift from a direct line of travel (single track) to a sideways motion of varying degrees in what is often called *two tracking*. With flexion of the back, such sideways locomotion can occur in the direction of flexion *(traver* or *renver)* or with the convex curve of the body leading (i.e., *shoulder-in*). Some degree of leg crossing occurs in such maneuvers.

Circling: Turning during locomotion, especially continuous tight turns *(circling)*, can involve leg crossing. Turns can be pivotal primarily around the forelegs or around the hindquarters (e.g., *pirouette*). In the *pirouette* the forelegs describe a larger circle than the hindlegs (Seunig 1956).

Piaffe and Passage: Leg motion can occur with little or no forward movement of the body. When a horse maintains a lofty, sustained trot-like action while remaining in place, showing springiness in its leg movements, it is called *piaffe*. When similar leg motions create a slow forward floating movement, it is referred to as the *passage* (the parade step of the ancient Greek horses).

Table 3.2: Motor Patterns, Postures, Emissions, and Other Behavior Patterns Characteristic of the Horse Ethogram

Alert	Head toss	Running walk
Approach	Head turn	Scratch
Attack	Hindleg lift	Shake
Avoid	Hindleg stretch	Shy
Back	Jump	Skin twitch
Balk (Jib)	Kick	Sleep
Ballotade	Kick threat	Slow gait
Bite	Knock	Smell
Bite threat	Lateral recumbency	Snaking
Blink	Levade	Snapping
Blow	Lick	Snore
Bolt	Lie down	Snort
Buck	Look	Squeal
Buck-jump	Marking	Stamp
Canter	Mezair	Stand
Capriole	Mount	Stare
Chase	Mutual groom (Allogroom)	Sternal recumbency
Chew	Nasogenital contact	Strike
Circle	Nasonasal contact	Strike threat
Copulate	Nibble	Suck
Cough	Nicker	Sunning
Courbette	Nip	Supplant
Crib	Nostrils flared	Swallow
Croupade	Nurse (Suckle)	Swim
Defecate	Pace	Tail depression
Drink	Parturition	Tail flagging
Drive	Passage	Tail raise
Drowsy	Pawing	Tail switching
Ears laid back	Pelvic thrust	Tongue manipulation
Ears lateral	Penis erection	Traver
Ears pricked	Penis extension	Trot
Ejaculation	Penis retraction	Two tracking
Eye roll	Piaffe	Upper lip movements
Flehmen	Pirouette	Urinate
Gallop	Prance	Wait
Get up	Present (Solicit)	Walk (diagonal)
Groan	Push	Walk (lateral)
Head extension	Rack	Weaving
Head flexion	Rear	Whinny
Head nodding	Renver	Windsucking
Head shake	Roll	Winking
Head stretch	Rub	Yawn

Swimming: Horses while *swimming* maintain leg movement in a sequence resembling the trot. The head is elevated keeping the eyes and nostrils above the water surface.

Jumping: Horses exhibit *jumping* over high elevated obstacles as well as over ditches and similar obstacles requiring broad jumps. In both cases, the forelegs are raised clear of the obstacle while the animal continues to propel forward with a final push by fully extending the hindlegs (Figure 3.4a). At take-off, the hindfeet are commonly at the site where the forelegs left the ground. The forelegs flex close to the chest as elevation is gained, extending subsequently to alight either simultaneously or sequentially as the hindlegs are momentarily flexed clear of the obstacle. The animal is fully off the ground during the jump. Occasionally while jumping, horses rotate the hindquarters to one side as the hindlegs reach maximum flexion. Although jumping can occur from most gaits, the running jump occurs usually from a canter or a moderate gallop.

Rearing Motions: Rearing is a motor pattern where the hindlegs remain on the ground while the forequarters raise high into the air (Figure 3.4b). A controlled movement where the forelegs are tightly flexed as the forequarters are moderately raised placing the spine 30-45° above horizontal is called the *levade* (Figure 3.4c). The weight is borne by the deeply flexed hindquarters. The greater this flexion the longer the horse can maintain the position. The *mezair* is a series of levade movements combined with forward motion accomplished by smooth jumps where the forelegs alight briefly followed by the abrupt alighting of the hindquarters.

An in-place leap or hop upward in a rearing-like attitude from the levade is called a *croupade* (Figure 3.4d), whereas a similar leap on the hindlegs where forward advance occurs is the *courbette* (Figure 3.4e). More than one leap may be induced before the forelegs again contact the ground. Such gymnastics are achieved by extensive development of muscle and coordination through training. Two other in-place leaps can be obtained from highly schooled horses. The *ballotade* is a high leap where the legs are flexed underneath with the hindlegs retracted as if ready for a kick (Figure 3.4f). In the *capriole* (Figure 3.4g) the hindlegs do kick posteriorly during the leap (Seunig 1956).

Bucking: A sudden humping or arching of the back with head and neck quickly lowered is called *bucking* (Figure 3.4h). Kicking with both hindlegs may follow. Frequently the horse also leaps or bounds clear of the ground, exhibiting what is called a *buck-jump* (Figure 3.4i). A horse usually performs these movements to rid itself of something on its back, as riders sometimes discover. A series of these motions can occur with leaps in erratic directions. Exuberant, playful horses while galloping at liberty sometimes exhibit bucking followed by a kick.

Kicking: Kicking with one or both hindlegs, while the forelegs remain in contact with the ground, is a common aggressive pattern of horses (Figure 3.5a). The suddenly flexed and elevated hindleg or legs are thrust quickly posteriorly as the weight is shifted over the forelegs. The neck may be lowered in the process. Two related motor patterns occur. One is *knocking* of the substrate (Figure 3.5b) with a hindleg (a similar raising and lowering of a foreleg is called *stamping*). The other is a *hindleg lift* used often by mares to block or bump away a foal, using the stifle to prevent access to the udder. Each of these movements can be forceful.

Striking: Striking is the often swift motion made by one or both forelegs in an anterior direction (Figure 3.5c) usually to hit or threaten another individual. Often it occurs with one leg while the other foreleg remains in contact with the ground. The neck is usually elevated. Striking can be done also by a horse during rearing.

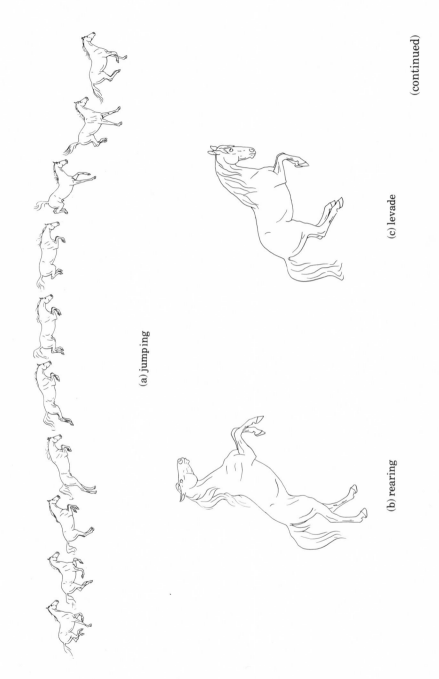

Figure 3.4: Jumping, rearing and leaping motor patterns of horses

(a) jumping

(b) rearing

(c) levade

(continued)

(d) croupade

(e) courbette

(f) ballotade

(continued)

(g) capriole

(h) buck

(i) buck-jump

Figure 3.5: Additional leg movements of horses: (a) kicking, (b) knocking, (c) striking, (d) pawing, (e) scratching.

Pawing: Pawing is similar to striking with a single leg except pawing is slower and the toe is dragged posteriorly in a digging or scraping motion. When used during investigation, the nose is usually oriented toward the substrate or object. Pawing is commonly repeated several times in succession. In addition to its use in scraping, Ödberg (1973) noted that pawing is occasionally exhibited as a displacement activity by horses restricted in forward locomotion. During such occasions, contact with the ground may be incomplete and the head and neck often remain elevated.

Scratching: Scratching with a hindfoot is often done by young horses and occasionally by adults, such as ponies. The body is flexed to one side and the hindleg on that side is extended forward so the hoof rubs the lowered head or neck (Figure 3.5e).

Pushing: Pushing is where a horse presses against something in an attempt to displace it. The head, shoulder, or thigh is used to push other organisms; the chest is often used to push against barriers.

Lying Down: Lying down is the process of going from a standing to a recumbent position. It is a continuum of motor patterns (Figure 3.6a) that may commence with the horse investigating the substrate with nostrils near the ground. Sniffing and trampling may ensue. If rolling is to occur, pawing of the substrate often takes place. Next the legs gather close together, often with a piaffe-like movement, with the head remaining low. Having positioned the legs, the forelegs begin to bend at the knees. As the forequarters slowly sink, the neck and head move forward. The body weight is concentrated on the somewhat flexed hindlegs. The forehand continues to sink, and the head is kept well forward. As the knees are about to contact the ground, the hindlegs fold, the neck elevates, and the body is abruptly lowered to the substrate. At this point, the horse is in *sternal recumbency,* resting somewhat on one side in such a way that the sternum and abdomen rest on the ground to either the right or left of the midline with head and neck upright. The legs remain flexed with only one hindleg extended free of the body.

To go completely prone to *lateral recumbency,* the horse rolls further onto its side and extends its legs while lowering the neck and head to the substrate. The upper foreleg is commonly anterior to the lower forelimb which is often slightly flexed at the carpal and fetlock joints (Littlejohn and Munro 1972). Either of the extended hindlegs can be slightly anterior to the other.

Rolling: Rolling is accomplished while recumbent by rotating onto the back with flexure of the legs (Figure 3.6b). The head and neck appear to assist in the effort to roll by providing leverage for the sudden body twist. The extent of the roll commonly stops along the back with muzzle pointing skyward; however, the animal may roll over completely onto its other side. If the latter occurs or the roll is inadequate, the horse often attempts to return to its back where it may rub against the substrate with legs thrashing as the back flexes laterally back and forth. After a few seconds, the horse returns to sternal recumbency.

Getting Up: The process of *getting up* onto the feet begins with the position of sternal recumbency. The weight is shifted posteriorly by elevating the neck as one foreleg then the other extend anteriorly lifting the forequarters clear of the substrate (Figure 3.6c). As one or both forelegs become stabilized, the neck lowers allowing the weight to shift anteriorly, and the hindquarters are raised by the hindlegs. On rare occasions, a foal varies the process and stands by first raising the hindquarters, reversing the sequence of lying down.

Shaking: Shaking is where the surface of the body as well as head and neck are rotated or vibrated rapidly. This frequently occurs after rolling. The entire animal vibrates momentarily casting away dust and other matter from the pelage. Localized quivering of the skin *(skin twitching)* occurs in response to localized stimulation of the skin, for example, by insects. Insects and other annoyances around the head and ears cause *head shaking.*

Rubbing: Rubbing can occur, for example, by a horse moving its lower jaw surface or muzzle against its forearm or by moving any part of the body back and forth or up and down against some object. *Licking* with the tongue and *nibbling* with the incisors are other motor patterns often directed at the pelage during grooming.

Mouth Movements: Biting motions are often directed at another horse by extending the head and neck while opening the mouth and directing the incisors at the other individual. If contact is made a *bite* occurs. The feigning of a bite without making contact is a *bite threat.*

Immature horses during submission occasionally display an up and down movement of the jaw while the lips are retracted at the corners of the mouth. This pattern has been called *snapping*, teeth clapping, Unterlegenheitsgebärde, and so on.

Figure 3.6: Motor patterns of (a) lying down to sternal and then lateral recumbency, (b) rolling, and (c) getting up.

Among the motor patterns involved in feeding are *upper lip movements* to separate and help lift food material, biting and cropping food with the incisors, use of the tongue to move the food into the mouth *(tongue manipulation)*, *chewing* the material with cheek teeth by crushing and lateral grinding motions of the lower jaw, and finally *swallowing*.

The *sucking* pattern of a foal is displayed by extending the head and usually elevating it above horizontal while protruding slightly the receptive tongue flattened against the lower incisors. Sucking readily occurs once the tongue makes contact with a teat or surrogate object. Occasionally neonates exhibit sucking in mid-air prior to successful nursing.

Head Movements: The head is capable of a variety of other motor patterns, some of which will be covered later under communicative behaviors. *Nodding* is the oscillatory movement of the neck in the vertical plane, causing the head the head to change elevation. *Head tossing* is a similar motion up and down but where head flexion and extension is primarily involved. *Weaving* involves relatively slow lateral motion of the head, neck, and forelegs repeated cyclically.

The behavior pattern called *flehmen* (Figure 3.7) is where the head is elevated and the upper lip is raised wrinkling the nose and exposing the gums. Such motor patterns occur in a variety of mammals, including most ungulates and felids (Schneider 1930, 1931, 1932a, 1932b, and 1934). In the horse, flehmen begins with extension and elevation of the head, usually after sniffing something. As the head approaches extreme extension, the upper lip is lifted maximally exposing the upper incisors and adjacent gums. The jaw is usually closed or nearly so. The ears and eyes generally rotate to the side (Dark 1975), and the third eyelid (nictitating membrane) appears as a whitish area covering the anterior portion of the eye. At its peak, the head is raised high above horizontal. In less than one minute, the head posture and facial features return to normal.

Figure 3.7: Sequence of the flehmen response. (Adapted from Dark 1975)

Horses also *yawn.* The yawn begins from a relaxed head position while either standing or recumbent. The mouth starts to open, and a deep inhalation occurs as the head is raised and extended. The eyes roll and close, at least somewhat, as the yawn reaches its peak (Figure 3.8). The elevated head sometimes turns and rotates slightly. The lower jaw may shift laterally when the mouth is wide open, and the otherwise relaxed ears may shift forward momentarily. Exhalation occurs quietly as the yawn regresses (Dark 1975).

Figure 3.8: Sequence of the yawn. (Adapted from Dark 1975)

Stretching a portion of the body occurs most often by either moving one hindleg *(hindleg stretch)* posteriorly fully extended, often while raising and lowering the back, or by elevating and sometimes extending the head (the *head stretch).* Such movements often occur after a yawn or a period of rest.

Eye and Ear Movements: Eye and eyelid movements occur. *Eye rolling* is where the eye rotates downward or posteroventrally and retracts exposing white scleral tissue above the pigmented iris. During eye rolling, the light colored nictitating membrane often moves over the anterior portion of the eye blocking some of the dark pigmentation of the iris. Although the nictitating membrane can be quickly raised and lowered, *blinking* normally involves the eyelids per se.

Ear movements are versatile and are controlled by the complex interaction of some fifteen auricular muscles (cf. Sisson and Grossman 1953). When the ears are *pricked,* they are forward in a vertical position with their opening directed forward. When the ears are *laid back,* their opening faces posteroventrally and collapses to some degree as they are pressed back against the upper part of the neck. The ears can be rotated individually to varying lateral positions between the fully pricked and the extreme laid back positions. The open portion of the ear rotates approximately 180° through this lateral arc.

The nostril openings can change diameter depending on the physiological and psychological state of the animal. *Flared nostrils* are those maximally dilated.

The attitude of lowered neck, extended head, laid back ears, bite threats, and forward locomotion while the neck slowly oscillates from side to side constitutes the behavior pattern called *snaking*. It is usually exhibited by stallions when attempting to drive or move other horses.

Tail Movements: Tail switching as well as *tail depression* (pressing the tail against the perineum) and *tail raising* occur by the interaction of five muscles. More will be said about motor patterns and postures of the tail and other parts of the body in subsequent chapters.

Part II

Behavioral
Development

4

Ontogeny of Behavior Patterns

PERINATAL DEVELOPMENT

At birth, a foal exhibits behavior patterns that have developed during the *in utero* period of approximately 340 days. From the third month onward, fetal movements can be detected which become more complex as gestation continues and the fetus matures. Bouts of activity and rest occur. The peak of fetal activity occurs about three days prior to parturition and appears to lead to the attainment of the birth posture (Fraser et al. 1975).

In free-roaming herds, births typically occur in late spring. Some foalings do occur at other times of the year, in all seasons. Yet, under most feral conditions foaling is rare during winter. Under management conditions winter births are not as uncommon when parturition can occur in a stall. In fact, to coincide with the custom of labeling a horse one year old at the beginning of the next January, some horse associations encourage foaling to occur early in the calendar year, out of phase with natural tendenices.

During parturition the forefeet of the foal appear shortly after the rupture of the chorio-allantoic membrane. At this time stimulation of the forelimbs may cause some motor response from the foal. But as the body of the foal passes through the maternal pelvis by additional uterine contractions, reactions by the foal cease, even to painful stimuli, until the hips are delivered (Rossdale 1967a). At this point straining by the mare ceases, and the foal is officially born.

Within seconds after the pelvic girdle of the foal leaves the maternal reproductive tract, the foal lifts its head and neck and assumes sternal recumbency. If intact, the amnion is thus ruptured, and breathing can commence unhindered by fetal membranes. The head is unsteady as the foal regulates its upright posture. The eyes are open. The ears remain back or protrude passively to the side. The tail is tucked down.

Newborn Thoroughbred foals weigh between 38-62 kg (84-137 lb), breathe at a rate of 65 ± 6.5 breaths per minute during the first minute of age, and have a rectal temperature of $37.1°$ to $38.9°C$ ($98.8°$-$102.0°F$) and a heart rate of approximately 69 beats per minute. During the process of trying to stand, the heart rate can be as high as 200 beats per minute before stabilizing at about 96 (double the adult rate). By one hour of age, the respiration frequency has dropped to 34 breaths per minute (as an adult it will be approximately 12). Body temperature averages about $38°C$ ($100°F$) after the first hour in healthy horses (Rossdale 1967b), 1968a, 1969).

In addition to the righting reflex shown in the first moments after birth, the foal's initial movements appear to be a reaction to restraint by fetal membranes and to the hindlegs being not yet free of the maternal reproductive tract. If the mare remains recumbent, crawling movements by the foal, using the anteriorly extended forelegs assisted by motions of the head and neck, cause the neonate to move away from the mare. These locomotor movements drag the foal's hindlimbs free of the mare's vagina and usually cause the umbilical cord to sever as the distance increases (Waring 1970a).

Movements of the foal continue, usually in bouts. The initial efforts to free itself from restraint appear to shift to attempts at getting up. Often at 15 minutes postpartum the foal has begun to raise its sternum off the substrate by pushing with forelegs extended anteriorly, maintaining its forehooves in contact with the substrate. The hindlegs during the initial efforts appear inert; nevertheless, repeated attempts to stand occur. Usually not until after another 30 minutes do the hindlegs finally flex sufficiently to assist in lifting the body free of the ground. If disturbed, the foal and the mare stand sooner than they would otherwise.

Meanwhile the eyes of the foal, accompanied by appropriate head movements, begin to show distinct binocular orientation by 25 minutes of age. Ten minutes after birth, Rossdale (1967a) was able to elicit the pupillary light reflex, and head jerking was caused by the flash of photographic bulbs. Auditory orientation becomes evident about 40 minutes postpartum when the ears begin to show distinct and independent orientation toward ambient sounds. Even before standing, the foal investigates its immediate surroundings using its eyes, ears, and nose. Periodic tactile and vocal stimulation by the mother begin soon after parturition (Waring 1970a).

Using data on 249 Thoroughbred foals, Rossdale (1967a) concluded the average time taken by neonates to stand was 57 minutes. The data ranged from 15 to 165 minutes, with more foals standing in the 40-60 minute interval than in any other period (Figure 4.1). Stop-motion film analysis (Waring 1970a) has shown the initial stance of the foal is unsteady with the legs spread laterally, hindlimbs extended posteriorly, and the forelegs positioned well forward with a slope of nearly $50°$. The crest of the neck is held with a slope of about $40°$ and the dorsal surface of the muzzle at $45°$. The foal shifts its neck and feet frequently to maintain its balance.

Locomotion forward, laterally, or backward is first accomplished by shuttling motions of the spread legs. These motions soon approximate a walking gait with little leg flexion. During the next hour, the leg flexion and walking pattern are perfected until coordination is achieved and the foal moves along easily.

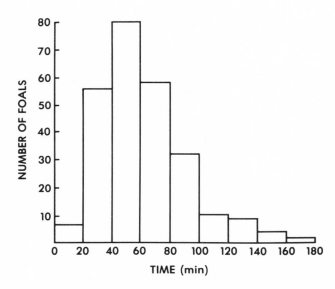

Figure 4.1: Time from birth to first standing for 249 Thoroughbred foals. (After Rossdale 1967a)

The sucking reflex can be induced within minutes after birth by objects in contact with the mouth. Tactile stimulation along the anterior half of the head triggers and maintains searching and sucking activity. When sucking has not yet been induced, I have observed standing as well as sternal recumbent foals exhibit spontaneous sucking motions in mid air 31-60 minutes postpartum. The lips and tongue were characteristically shaped and sucking sounds could be heard; the head was extended and swayed from side to side as the mouth was elevated.

Nosing, smelling, and licking of nearby objects occurs during the foal's prenursing investigations. For example, the mare's forearm, girth, gaskin, and perianal region are thus investigated if contacted.

Successful nursing is dependent upon the mare's willingness to stand motionless and the foal's ability first to stand and then to carry out nipple searching activities. Some mares subtly position themselves in a way that all the foal needs to do is extend its head and begin sucking. Such fortunate foals nurse soon after standing. Often foals inadvertently delay nursing by searching for long periods around the mare's forelegs. In other cases, the restless mare may move away each time the foal probes the apparently tender udder region. In the latter instances, human attendants often intervene to restrain the mare and guide the foal to the milk source. Rossdale (1967a) found that foals born in box stalls nursed between 35-420 minutes following birth. The average was 111 minutes (Figure 4.2). Attendants facilitated some of the initial nursing bouts; yet, other studies of confined and free-ranging horses have found similar results (e.g., Tyler 1969, Waring 1970a, Boyd 1980).

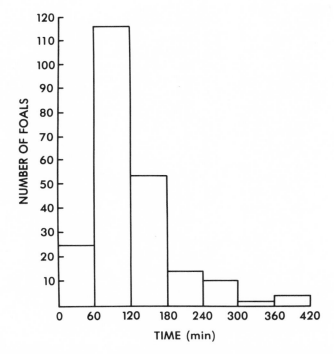

Figure 4.2: The time of first nursing from the mare. Data from 245 foals. (After Rossdale 1967a.)

Once nursing has occurred successfully the first time, the foal returns progressively more easily to the flank area and teats when attempting subsequent nursings. The nursing intervals generally vary between 10 and 90 minutes for the first 24 hours. Drummond et al. (1973) found that gnotobiotic foals fed ad libitum with milk formulated to approximate mare's milk drank 300-400 ml per feeding.

Defecation may occur prior to one hour of age in foals standing successfully; urination follows a few hours later with posture typical of the sex of the foal. Defecation occurs with the foal spreading the hindlegs, raising the tail 40° or higher above horizontal, and depressing the croup protruding the anal area posteriorly. Straining in an attempt to pass firm pellets is not uncommon.

The first few attempts to lie down often end in rough collapses, although the foal during the second hour postpartum may try to slowly flex its closely placed legs to go down steadily first to its knees. Unsuccessful attempts to go down are often made at this early age, only to return instead to standing or walking. Resting may eventually be done by fatigued foals while standing. Not until after several tries does a foal lie down with coordination and ease. Having once gotten to its feet unassisted, subsequent standing is usually done readily and with success.

Vocalization is rare from newborn foals. Weak whinnies and squeals may be emitted during the first hour by foals when distressed or restless. Yet it is during the second hour postpartum that the mare and foal overtly respond to the other's sounds. The mare is the more vocal member of the pair.

At the end of one hour of age, the foal shows at least basic abilities in righting itself, maintaining its posture, investigative behavior, standing and moving about, care-seeking behavior, agonistic withdrawal when restrained, and sometimes other behaviors, such as ingestion, vocalization, and defecation (Figure 4.3).

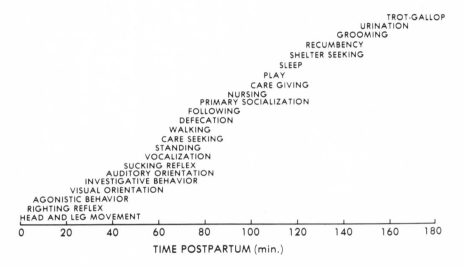

Figure 4.3: Progression in the onset of behavior patterns of neonatal foals. (Adapted from Waring 1970a and Reed 1980)

During the second hour postpartum, the foal begins to follow and remain close to the mare, nuzzles her, and seeks her side upon the approach of others. The foal seems to show concern for the mare when she struggles with discomfort, such as when trying to expel the placenta and fetal membranes. When the mare is down and exhibits discomfort, the foal may circle her restlessly and may whinny loudly following her groans. These behaviors are used as indicators that primary socialization is occurring at this early age (Waring 1970a, 1970b).

Sleep behavior begins as brief naps in the second hour of age and progresses thereafter until sleep and drowsiness occupy the majority of the foal's early life. Sleeping commonly occurs while completely prone or in a relaxed sternal recumbency, but when unable to lie down a foal will stand and doze with eyes mostly closed and neck nearly horizontal.

Fear of new objects begins as early as the end of the second hour of age; but with the security provided by the close proximity of the mare, the foal continues to investigate its surroundings. Foals resist restraint from the early minutes of age; however, learning to adapt to restraint-type handling can occur in these early hours. Foals receiving such early handling separate from their mothers to greater distances and show more self-confidence in exploratory behavior. They also tolerate restraint better when older (Waring 1970b).

At the end of two hours of age, the typical foal has perfected its earlier abilities until it can now walk easily, nurse, follow its mother, vocalize, interact socially with the mother, and seek shelter beside her. Fear and sleep have also appeared.

After several more hours the foal can, in addition, combat insects by nipping at its side and also by moving its tail and legs. It urinates typical of its sex, and it can trot and gallop with ease. It shows brief spells of exuberant play and has begun mouthing various objects, such as hay, grass, twigs, and feces. Some ingestion of these solids may occur. Tyler (1969) once observed a newborn nibble grass for a total of 15 minutes while the mare struggled for 40 minutes to expel the afterbirth. Foals have also been observed to exhibit in the first 24 hours rolling, scratching, rubbing, and the facial expressions of flehmen, yawn, and submissive snapping (Unterlegenheitsgebärde). Swimming, too, is possible. Ron Keiper witnessed a day-old foal swim a four-foot-deep tidal stream to keep up with its mother (Ford and Keiper 1979).

POST-NATAL DEVELOPMENT

The foal's behavior begins to exhibit greater rhythmicity after the initial perinatal period. Nursing intervals, for example, become more regular; yet like many other behavioral characteristics, nursing too changes with age. During the first week, Tyler (1969) noted the frequency of nursing during the day was approximately four bouts per hour. Thereafter the frequency decreased, as did the duration over the first few weeks (Figure 4.4). By the sixth week, foals nursed on the average twice per hour; by the fifth month, the frequency decreased to once each hour. Feist and McCullough (1975) reported similar nursing rates for feral horses.

Not just the foal regulates nursing, but the mare also encourages or discourages nursing by her activity and posturing. Some foals in box stalls have been observed to routinely nurse while standing along a particular side of the mare (showing individual preference for either the right or left side); yet in pasture the same mare-foal pairs do not exhibit such position effect (Waring 1978).

Grazing behavior in foals is infrequent during the first week, but time spent grazing increases gradually during the next few months. Tyler (1969) observed a more rapid increase in the time spent grazing after foals reached four months of age (Figure 4.5). She found grazing time was significantly higher in the late afternoon for free-ranging New Forest ponies than for either the early morning or mid-day time period (see Table 4.1). At four months the foals grazed an average of 16.3 minutes per hour during daylight; by twelve months, grazing occupied an average of 44.4 minutes of each hour. Similar changes were noted in Camargue foals by Boy and Duncan (1979).

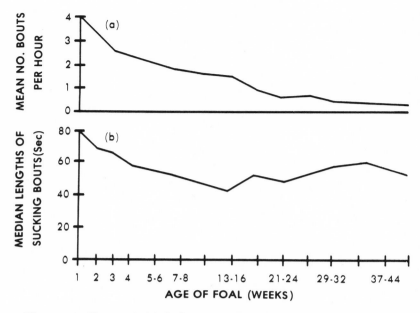

Figure 4.4: Changes in (a) the frequency of nursing and (b) the length of sucking bouts as foals mature. (After Tyler 1969)

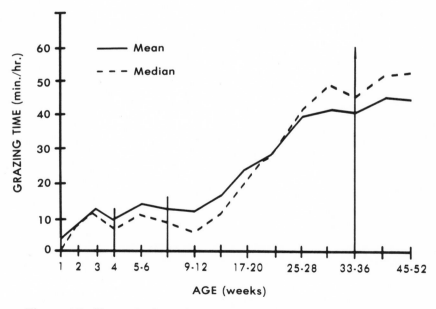

Figure 4.5: Change in the proportion of time foals spend grazing prior to weaning. (After Tyler 1969)

Table 4.1: Minutes per Hour New Forest Pony Foals Were Observed Grazing During Morning, Noon, and Afternoon Periods*

Age (weeks)	0600-1000	1000-1400	1400-1800
1-2	6.5	5.1	7.3
3-4	9.8	11.7	13.2
5-6	15.2	14.4	16.4
7-8	6.7	17.7	13.5
9-12	10.4	13.2	17.0
13-16	13.7	11.3	23.1
17-20	17.5	21.3	32.6
21-24	19.3	20.4	34.9
25-28	25.5	41.1	34.2
29-32	44.3	37.3	41.7

*Data from Tyler 1969.

During the first four months, foals spend about half of each hour resting, primarily while recumbent (Figure 4.6). Resting tends to be distributed throughout the day. On the open range of the New Forest in England, Tyler (1969) noted that during the first two months, foals were recumbent 70 to 80% of their total resting time. Subsequent to three months of age, foals spent less and less time resting; so that by nine months of age, resting by foals was not seen in about half of the hours of observation (Figure 4.7). Resting time was minimal during late afternoon. In the Camargue of southern France, Boy and Duncan (1979) reported lateral recumbency decreased from 15% in newborn foals to 2.7% in nine-month-old pre-weanlings; sternal recumbency decreased from 17.9% to 13.2%; and resting while standing increased from 8.1% to 11.8%.

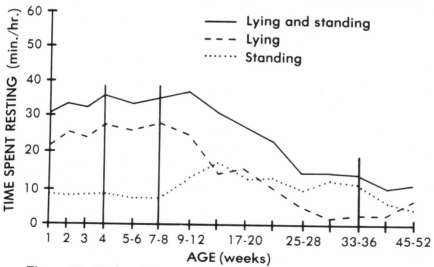

Figure 4.6: Change with age in the time spent resting by foals during daylight hours. (After Tyler 1969)

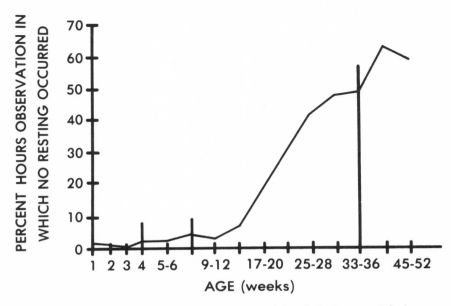

Figure 4.7: Shift with age in the proportion of time foals do not rest during daylight hours. (After Tyler 1969)

Initially the urination frequency is high in young foals; the defecation rate is just the opposite (Figure 4.8). By seven months, the hourly urination rate gradually shifts to the adult mare rate of approximately once every four hours. Defecation proceeds from about once every ten hours during the first week to nearly once every 3 to 4 hours by five months of age (Tyler 1969). Health and diet can influence such rates.

Up to the age of 3 to 4 weeks, coprophagy (eating feces) is common in foals. Neonates as well as older foals can occasionally be seen eating pellets. Generally the fecal material is that of the mother, less often that of the foal. In most cases, rather than show ingestion, foals simply sniff fecal piles. Pawing of the pile may proceed the small amount of coprophagia. Foals rarely defecate onto other feces; however, as foals become older urination onto feces becomes a rather common response of both colts and fillies (Tyler 1969).

As the skin becomes irritated by biting insects and other causes, foals spend long periods grooming themselves. Scratching of the head and neck with a hind foot as well as nibbling the legs, rump, and back are commonly seen. Coordination is developed over several days following birth before balance remains steady. Rubbing, shaking, and tail switching are also utilized. Foals as young as the first day of age sometimes discover the benefits of rubbing to relieve skin irritation and subsequently spend bouts of up to 15 minutes rubbing selected objects in their environment.

During its first week, a foal may begin to interact in mutual grooming. The mother or other foals are commonly involved, occasionally other group members with the exception of the dominant stallion. After four weeks of age, foals spend more and more time mutually grooming with other foals. Nibbling of the neck, mane, withers, or forelegs is done by the grooming pair. If bouts continue, the hindquarters receive attention. In the study by Tyler (1969), mutual grooming among foals reached a peak in frequency when foals were 3 to 4 months old. Bouts usually lasted only a few minutes.

Typically the relationship between the foal and its mother is reinforced over the first few days, and the distance between the pair remains slight. Curious group members, e.g., yearlings, are threatened away by the mare. The innate tendency of the foal to follow large objects solidifies into a strong social attachment between the foal and its mother. Rarely, a foal's attachment may be directed toward an inappropriate large object in the environment, such as a tree (cf. Tyler 1972). In such a case, the mare may abandon the foal; yet if the foal's fixation is redirected to the mare early enough, a successful mother-foal relationship can eventually develop.

Husbandry and handling practices may greatly affect the mare-foal relationship and the social development of the foal. To the extreme, a foal could be raised on a mechanical milk dispenser isolated from all other organisms, or a foal could be raised in isolation from other horses with only human attention and social contact. Grzimek (1949a) reared a foal under the latter conditions during the foal's first two months. When first confronted with other horses at 64 days of age, the foal exhibited fear, actively avoided the other horses, and attempted to remain with human handlers. The foal did not view horses as conspecifics. A similar defect of normal social behavior has been seen with isolation-reared foals with milk dispensed mechanically (Williams 1974).

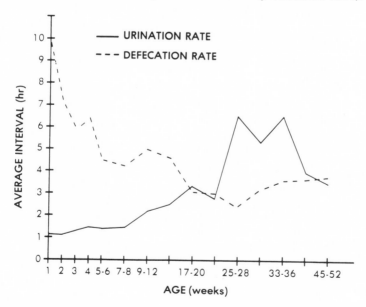

Figure 4.8: Frequency of urination and defecation in foals. (Data from Tyler 1969)

Most owners do not desire their horses to be completely human oriented, but the other extreme where a horse fears human contact is also seldom desired. A middle ground is usually sought. Some of my early work with foals was directed at developing human-socialized animals that also maintained normal social development with the mother and eventually with other horses. We began by utilizing a handling routine developed by Gertrude Hendrix (Marwick 1967) which commenced as soon as the foal went down subsequent to its first nursing. It soon became apparent that primary socialization to the mother was beginning much earlier and that in order to achieve concomitant socialization to humans, our efforts would need to commence sooner. We varied treatments between foals; some received no human exposure, others received active handling and fondling in their first hour or two, others received passive exposure to humans by a person quietly sitting in the foaling stall, and still others were exposed to a human mannequin standing in their stall. We concluded from this preliminary work that primary social attachment is dependent not only upon initial exposure during the early sensitive period but upon propinquity over time. Continued association maintains and strengthens the bond. It typically occurs between the foal and mother. If human socialization commences through early handling, the social attachment fades when direct human interaction is rare compared to interactions with a companion such as the mother.

The foal-mare relationship can be maintained, even with human handling, provided the pair is not separated for prolonged periods. I have isolated foals from 5 to 70 minutes of age with no permanent disruption of the relationship between mother and young. When illness strikes the mare or the neonate, such as the convulsive syndrome of foals (Rossdale 1968b), the pair bond may appear disrupted; yet when the pair is kept together, development of the relationship usually proceeds satisfactorily if health can soon be restored. The drive to establish an intimate relationship is not readily lost.

Successful early bonding appears important in the occasional fostering cases that have succeeded with foals as old as 3 months of age (Rossdale 1968b, Tyler 1969). In such cases, mares who have recently lost their own foal can be induced to accept recently orphaned foals. The foals too are receptive once a social void exists.

Early handled foals when compared to unhandled foals show more exploratory behavior and attenuate fear responses more readily. They move away from their mothers more readily and to greater distances when first turned outdoors, approach other organisms, and in general show more self-confidence. Such activities cause the mother to spend more time following her foal and herding it away from contacts with others. Unhandled foals are reluctant to leave the side of their mother during their initial exposure outdoors. Therefore, early handled foals can be subject to dangers resulting from their zealous curiosity (Waring 1970b, 1972).

The intimate two-way bond that normally develops between the neonate and its mother gradually changes with the foal's increasing age. The widening of the distance between the pair is one indication of the relaxation of the intimate relationship. In the first week, foals with normal early experience spend more than 90% of their time within less than 5 meters of the mare. By the fifth month they spend about half their time in such close proximity; and by the eighth month, foals are within 5 meters of the mother only about 20% of the day (Tyler 1969).

As the relationship between mother and foal changes, the foal proceeds to develop a relationship with another foal or with a yearling. Progressively more time is spent with the new companion. In a sibling foal-yearling relationship, the foal's independence generally progresses more rapidly than if the foal were associated only with the mother. And the yearling sibling may remain in closer proximity to the mother than it would otherwise, because of its association with the foal. The relationship between mother and offspring, although never as intimate as when the foal was small, is normally maintained to some extent into the offspring's adulthood. The offspring especially exhibits periodic interest in grooming and associating with the mother.

The pattern of play behavior also changes with age and companionship. Beginning in its first day, the neonate exhibits periods of exaggerated and often incomplete motor patterns that are considered play activities. Play behavior of foals may contain components of locomotor, agonistic, sexual, ingestive, grooming, and other behavior patterns. Exaggerated withdrawal and approach (e.g., galloping play) can be seen within a few hours after parturition. Biting and nipping at the mare's legs, tail, and other parts of the body are also characteristic of the foal's early play. Initially the mother is the focus of the foal's play behavior or else solitary play occurs; but as the foal establishes new social relationships, play focuses increasingly on the new foal or yearling companions. Peer relationships thus develop.

Vocalizations and expressive movements are observed in young foals, but no specific ontogenetic pattern has been recognized. Nevertheless, one expressive movement called snapping or Unterlegenheitsgebärde is characteristic of young horses, apparently as a submissive expression. The behavior pattern consists primarily of vertical jaw movement while the lips mostly cover the teeth and the corners of the mouth are drawn back. The display is relatively silent. The expression occurs when the immature horse appears fearful as it approaches or is approached by another horse or large object. Table 4.2 shows the variation in the frequency of snapping as well as the recipients of the expressive movement as age increases. Snapping rarely occurs after the second year. No difference in frequency of snapping has been found between colt and filly foals; however, colts snap more to stallions. Williams (1974) noted that foals reared on a milk dispensing machine and isolated from other horses showed snapping at the approach of strange humans but not toward familiar ones. The frequency toward humans was noticeably higher in such foals than in foals reared from the beginning with their dam or with other orphan foals.

Sexual behavior develops gradually in young horses. Mounting behavior can be exhibited by both colts and fillies during their first to fourth week. Mounting is more frequently exhibited by colts, and subsequent to four weeks it is exclusively a male characteristic. At first, foals incorrectly orient along the side or neck of the recipient (usually the mother) as mounts are attempted. However, even before the end of the first week, it is unusual for a foal to mount incorrectly. Neither penile erection nor pelvic thrusts occur during these early mounts.

Table 4.2: Total Number of Snapping (Unterlegenheitsgebärde) Responses Exhibited by Female Ponies up to Four Years of Age

			Recipient of Responses		
Initiator	Foal	Yearling	Adult Female	Adult Male	Total
Females					
Foal	5	25	92	25	147
Yearling*		1	42	38	81
2 year old			7	6	13
3 year old			5	6	11
4 year old					0
Total	5	26	146	75	252

*In this study, twice as many filly foals were present as filly yearlings.
(Data from Tyler 1969)

By two months of age, colts can be seen with full erections as they rest or interact with other horses in grooming or play. They may investigate the urine and genital region of estrous mares but usually exhibit no further interest. Tyler (1969) witnessed an exceptional case where a 3-month-old colt briefly mounted a 2-year-old filly in estrus; intromission was unsuccessful because of the colt's small height. As colts reach the age of two, greater attention is given to estrous mares. The age New Forest pony colts first achieved copulation varied in Tyler's study from 15 months to nearly 3 years.

Unlike males, fillies show little sexual behavior until they reach puberty and first exhibit estrus. They then approach males, present to them, and urinate frequently, much like older mares. Fillies can come into estrus as early as their second summer when 14 to 17 months of age, but conception is very low as yearlings. Stallions tend to ignore very young mares (Feist 1971, Tyler 1969). And although colts may copulate with estrous fillies, their reproductive development is incomplete.

Like sexual behavior, weaning varies with the foal's situation. Under intensive management, foals are often weaned and separated from their mothers at about six months of age. Yet, unlike the situation at most horse management facilities, in free-ranging herds weaning occurs when foals are almost a year of age or even later. Tyler (1969) observed most mares weaned their offspring only a few weeks or even just days before the next foal was born. In many cases, weaning is abrupt; the mare suddenly begins to threaten and avoid her young whenever it approaches to nurse. Mares which do not give birth to a new foal show little observable change toward their previous foal; nursing may continue through the second summer. In most cases, weaning occurs before the next spring as 2- and 3-year-olds are rarely seen to nurse. Occasionally some mares with new foals allow the previous young to continue its nursing behavior. One offspring may nurse from the side and the other reach a teat from between the mare's hindlegs.

Although foals may be receptive, it is rare that a mare will allow a foal other than her own to nurse.

5

Play

Play behavior appears to have a major role in the behavioral, social, and physiological development of equids, as in many other mammals. In horses, play includes such activities as (1) solitary or group running with exaggerated motor patterns, (2) approach-withdrawal patterns such as alternate chasing, nipping, and pushing as well as (3) the tossing or manipulating of objects by mouth. Playful activities have components seen in other behavioral patterns; yet, the lack of seriousness and incomplete sequences usually make play distinctive.

Although play is characteristic of young horses, mature animals also occasionally play. However, Feist (1971) noticed dominant males of feral bands usually curtailed vigorous locomotor and social play occurring among adult members of their social unit.

With the exception of play between foals, locomotor and social play are normally restricted to horses within the same social group. Occasionally foals from different social groups interact in playful activities as their bands are nearby. Play is greatly reduced during periods of extreme ambient temperatures, food scarcity, and most other occasions of physical and physiological stress.

SOLITARY PLAY

Play activities that occur without an interaction with other organisms can be considered solitary play. In horses, solitary play activity is primarily either a type of locomotor play or some form of manipulative play.

Within a few hours of birth, foals have typically begun vigorous locomotor play (Figure 5.1a). They move in a frisky manner to and from their mother or move in small circles exhibiting galloping, swerving, bucking, jumping, striking, and kicking. Similar activities are seen in Przewalski foals (Dobroruka 1961). The dashes are intially limited to within a few meters of the mother or some other center of the foal's early environment. These exuberant bouts of playful activity can last a few seconds or up to several minutes before the foal again becomes more quiescent. Play provides most of the vigorous exercise in foal development, at least in the first six weeks (Fagen and George 1977).

Figure 5.1: Examples of solitary play in horses: (a) locomotor play of a foal, (b) manipulative play of an adult gelding.

As the foal develops, the distance covered during locomotor play increases. At this stage in the foal's development, other young horses typically become play companions and solitary play becomes increasingly uncommon provided social contacts are feasible.

Manipulative play also appears early in the behavioral development of foals and can be seen occasionally in adult horses. Foals as young as two hours of age can be seen periodically nibbling, biting, or pulling at objects in their environment. Sometimes they lift the item, but often their exaggerated movements are an incomplete sequence; so that after brief contact, the foal shifts to other motor patterns. Approach-withdrawal movements often accompany the playful biting. Pawing of the object may also occur.

Horses, especially stabled animals, can occasionally be observered picking up sticks, boards, rags, pieces of paper, buckets, and other objects and swinging or tossing them. These individuals usually repeat the act several times in one bout. In some cases, the objects appear to be maneuvered toward other horses in the vicinity (Figure 5.1b).

Some stabled horses are notorious for their ability to manipulate electric light switches, door latches, and other devices within reach of their stalls. Such activities commonly appear to be a form of solitary play and stimulation much to the chagrin of horse owners who find motorized barn doors have been opened, a light switch has been activated, or a horse has opened its stall door and has been inspecting the barn.

PLAY BETWEEN FOALS AND THEIR MOTHERS

Play between young foals and their mothers is usually a situation of the mothers quietly enduring the playful activities of their foals. The mothers in such interactions tolerate the nibbling, biting, pawing, kicking, and other antics of their offspring and seldom exhibit play behavior themselves. Yet the mother is almost always the focus of the neonate's play.

Within a few hours of birth, foals show an interest in nibbling and poking parts of their mother's body. Neonates can be seen to playfully bite at the legs and sides of their mother. They pull and chew on the mother's mane and tail. Bouts of exuberant galloping occur around as well as to and from the mare. At times a foal in its enthusiasm may strike, kick, or mount the mother or even attempt to jump her while she is recumbent.

Although the mother is initially the center of the foal's play behavior, in time the focus of play shifts to peer companionship (Figure 5.2). Tyler (1969) found the percentage of hours of observation where foals played with their mothers or on their own decreased from 56% in the week postpartum to only 7.4% in the seventh and eighth weeks. Conversely, play with other foals or with yearlings could be seen to steadily increase over the same period.

By the time foals are about two weeks of age, the exuberant and sometimes rough biting of the foal toward the mare becomes more gentle, and the mare begins to respond with nibbling of the foal's body. Mutual grooming thus ensues (Tyler 1969).

PLAY BETWEEN FOALS AND OTHER YOUNG

On open range, few interactions occur between foals or between foals and older immatures until after two weeks of age. Initial interactions are usually visual investigations, and eventually the animals touch the other's muzzle before swiftly returning to their mother's side. In 158 hours of observation of second week old foals, Tyler (1969) found play or related interactions with foals or yearlings occurred in only 6.3% of the hours.

After the third week, playful interactions between foals and other young horses become more numerous. Approach, sniffing, touching, nibbling, grooming, threatening, kicking, withdrawal, and exuberant galloping with bucking and rearing are among the types of activities. Some foals paw repeatedly at other foals until they stand and become play companions (Tyler 1969).

During the first month, there is little difference in play behavior between fillies and colts, except in mounting frequency. Young colts mount their mothers or peers more often than do fillies. For example, Tyler (1969) recorded colts in their first month mounted approximately once every 5 hours of observation, whereas fillies mounted only once in 37 hours.

Subsequent to the first month of age, play of colts differs markedly from that of fillies. Colts as pairs spend long periods play fighting (Tyler 1969). Such aggressive play occurs between colts of similar age and also between foals and yearlings. Pairing is typical, and thereafter the partners seldom interact directly with other colts except in chasing.

In play fighting, each colt attempts to bite the head and neck of his opponent and push the opponent off balance. They rear and strike at each other and bite at the forelegs often causing the opponent to drop to his knees. They also may bite at the other's hindlegs, causing circling to occur. Effusive galloping with chasing, rump biting, and kicking may occur before another bout of head-to-head interactions. Although rough, the play fights are not vicious. The pair frequently exhibits mutual grooming between bouts of aggressive play. In 28% of the play bouts among foals observed by Schoen et al. (1976), the bouts were interrupted by both foals exhibiting a head and facial display that resembled flehmen with the ears laid back.

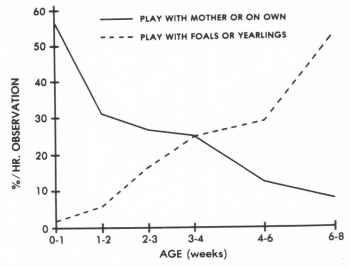

Figure 5.2: Play activity of foals and their choice of play partners during the first eight weeks. (After Tyler 1969)

Occasionally as foals, a colt and filly in the same social unit will become play and mutual grooming companions. Each shows a preference to interact with the other and for most activities they ignore other peers. Play fighting is evident in such pairs but is usually not as rough as between two colts or between adult males pastured together (Figure 5.3).

Figure 5.3: Play fighting among males.

Male foals without filly companions periodically approach female foals for mutual grooming. Nibbling and grooming may ensue. Yet, whenever colts begin to get rough and invite play fighting or mount, the female foals threaten them and try to avoid their biting and playful mounting. The fillies lay back their ears, bite, and kick.

Playful interactions between fillies are relatively uncommon compared to similar interactions between colts or between colts and fillies. Perhaps, as Tyler (1969) has suggested, it may be the precocious sexual nature of colts that leads to more play when males are involved. Sexual elements are evident in play between foal colts and fillies, where the male nibbles the hindlegs and rump of the filly and attempts mounting. Of the 423 playful interactions Tyler observed between free-ranging foals, half involved a colt and a filly, 34% involved two colts, and only 16% involved two fillies.

Play between fillies is mostly locomotor play, where one foal approaches or moves away from another in a frisky manner using exaggerated movements or they gallop side by side. Chases occasionally occur. Mutual grooming is the common form of interaction between fillies.

Wells and Goldschmidt-Rothschild (1979) confirmed that play is mostly among peers and that colts play more than fillies. They observed that during play fighting hindquarter threats occurred considerably more often than head threats. As play bouts ended, they noticed a tendency for the subordinate horse to give the last rear threat and the dominant partner to be the last individual to be the chaser.

PLAY BETWEEN YOUNG AND ADULT HORSES

Play between foals and adult mares other than their own mother is rare. Most mares threaten away foals that are not their own. Yet siblings and young mares are more tolerant of the playful biting and rearing of foals. They often passively allow foals to play just as the foal's mother tolerates the playful antics of her young.

Colts occasionally show particular interest in young mares and exhibit mounting and sexual interest. Under these circumstances, especially when the mare is in estrus, his behavior no longer appears to be play but true sexual behavior. Tyler (1969) observed a 3-month-old colt show such behavior to a receptive 2-year-old mare. Although intromission was attempted, it was not successful.

Stallions and geldings are submissively approached by young horses and often tolerate their playful behavior when focused on them. Usually the snapping display is exhibited by the foal or yearling during the approach. The young horse may be allowed to nibble the male's legs and tail or to nuzzle the adult's head or sheath. Tyler (1969) noted that 76% of such interactions involved young colts; 24% involved fillies. When stallions threatened approaching foals, the foals exhibited further snapping or withdrew. Rolling over to dorsal recumbency with legs uppermost was an additional form of submissiveness observed in Camargue foals (Kiley, cited by Tyler 1969).

Colt foals and yearlings occasionally play fight with adult males who gently frolic with the youngsters. The stallion ends such activities by walking away. The colt often follows the stallion inviting renewed play by rearing and pulling at the adult's mane (Tyler 1969). Although stallions may tolerate the presence of 2-year-old males, by the age of 3 years males begin to be treated like rivals.

Another type of social play that is often seen among horses is the exuberant locomotor activity that occurs prior, during, or soon after a refreshing storm or upon release from confinement. As one horse begins the playful frolicking, companions tend to join the activity. Galloping, rearing, kicking, circling, and other vigorous exercise thus briefly occur before the group returns to more quiescent activities.

6

Investigative Behavior

Investigative behavior facilitates the behavioral development of horses by exposing the animal to new objects, varied environmental situations, and new experiences. It permits the horse to become aware of its environment, not only to avoid hazards but also to learn traits important for its various biological activities. For example, through investigative behavior the animal finds potential danger, food and water, social companions, comfortable resting sites, and pathways. Throughout much of each day, horses exhibit investigative behavior, often while in other types of activity.

By the end of the first half hour postpartum, foals frequently exhibit visual investigation of their surroundings using monocular as well as binocular vision. While still in sternal recumbency, the foal rotates its head and eyes looking around and often fixes its gaze on nearby objects. During the second half hour following birth, the ears of the foal begin to independently rotate to investigate environmental sounds. By this time olfactory, tactile, and possibly gustatory senses have also commenced and are used in the pre-nursing investigative activity of the foal. Once the foal is standing, it moves cautiously nosing, smelling, and licking objects in its immediate vicinity. Objects at or just above head height are especially explored, such as the mare's forearm, girth, flank, gaskin, and perigenital region as well as tree trunks or stall walls. At this stage, contact along the dorsal part of the muzzle induces the sucking reflex, and the foal appears highly motivated to nurse.

Once the foal has successfully nursed, the motivation of most of its subsequent investigative behavior seems more directed at environmental awareness than nursing. The foal investigates its mother's body and the surroundings. It may nibble grass, straw, or fecal material. Each new object in the immediate vicinity is visually, auditorily, tactilely, olfactorily, and sometimes gustatorily investigated by extending the head, muzzle and occasionally the tongue. Movements are generally slow but jerky.

As the foal grows and as positive experiences accumulate, the foal's realm of exploration also increases. Negative experiences, where the foal has been hurt or frightened, cause the foal to be hesitant to experience similar events, and investigative activity may be temporarily inhibited. Yet with each favorable experience, a foal seems eager to explore new aspects of its environment.

The mother usually limits the foals early social contacts and range of exploration. Once the foal commences peer play activities, the opportunity for furthering environmental exploration and broadening its experiences are greater. Foals which experience neonatal handling and halter training tend to exhibit far greater exploratory interest and confidence than their unhandled neonatal peers (Waring 1972).

Throughout their lifetime horses continue to become alert to new objects; sounds as well as odors are, at least momentarily, investigated. A horse may orient its head in the direction of the stimulus, whereby the ears and eyes are directed forward. If the stimulus is at a distance, the neck is usually raised, elevating the head. If the stimulus is nearby, the head may be flexed in a collected position for visual scrutiny or extended and the neck lowered enabling the horse to smell and possibly touch the object.

Investigative activity can occur without the horse orienting its head directly toward the stimulus source. For example, stimulation from one side of the horse may only cause the eye and ear on that side to rotate and investigate. If minor stimulation is behind the horse, the ears and eyes typically rotate in that direction without the head or body becoming reoriented (Figure 6.1). The more suspicious the stimulus the more the horse tends to orient its head and body toward the stimulation. Alertness by one member of a group often induces similar behavior in other group members.

Horses continue to investigate new objects, intruders, sounds and odors until they appear to have determined if the stimulus requires additional action, such as flight or some social interaction. In most cases, the initially attentive horse returns to its previous activity subsequent to the brief investigation of the stimulus. Often the investigative response is only a subtle eye turn or ear movement. At other times, especially with novel stimuli, the alertness and investigative response of the horse is overt and unmistakable (Figure 6.2).

If new objects in the environment do not induce flight, they are generally investigated at close range, often by circling the object. Zeeb (1963) found that the Dülmen horses he observed would approach and investigate a motionless human, but they maintained a distance of 3 to 5 meters when the person walked near them. The horses withdrew and would not approach a person moving like a quadraped with hands and feet contacting the ground.

Excrement of other horses is frequently investigated by young and adult horses of both sexes. The neck is lowered and head extended permitting contact and smelling of the site. Frequently a flehmen response is given where the head elevates and extends while the anterior portion of the upper lip curls dorsally. Following the investigation of fresh feces and urine, the investigating horse moves over the site and oftentimes adds its own excrement to the site before departing (Feist and McCullough 1976).

When one horse investigates another horse, it often approaches with neck elevated while head, eyes, and ears orient toward the recipient. At other times, the investigating horse circuitously approaches utilizing monocular vision. If both horses participate, naso-nasal contact is common, accompanied by smelling and audible exhalation. Generally other regions of the body are then investigated, such as the head and neck. If only one individual is motivated to investigate, it may concentrate on the flank or perianal region of the other horse. If neither horse becomes aggressive, they often remain near each other until distracted.

Horses exhibit similar investigative and approach behavior toward life-like models of horses and full-sized two-dimensional horse sketches (Grzimek 1943a). The more such test objects deviate from being horse-like in body form, the less horses respond as if to conspecifics.

When a horse is in an approach-withdrawal situation, fear can prevent or impede close investigation. Avoidance is typical. Thus an anxious horse may be repelled by slight or even imaginary barriers, such as open doorways or water; whereas, when calm the same horse may approach, investigate, and proceed without incidence.

Figure 6.1: Subtle visual investigative response of a horse toward photographer while continuing to graze. (Photo courtesy of R.R. Keiper)

Figure 6.2: Overt alertness and investigative responses of horses.

Part III

Maintenance Activities

7

Resting and Sleep

The daily resting cycle of horses is polyphasic, that is, with more than one period of quiescence occurring per 24-hour period. Sleep may occur in several of the rest periods. In foals, brief naps often commence in the second hour after birth. Resting bouts continue to occupy more than half of a foal's time until about 3 months of age; the frequency then begins to decrease (Tyler 1969). For most resting bouts, young foals become recumbent; yet, after 5 months of age, standing becomes the more common resting posture, at least during daylight hours (see Figure 4.6). Nevertheless, young horses continue to rest in sternal or lateral recumbency more than adults.

Adult horses frequently rest while in a standing position. The so-called stay apparatus of the limbs (involving various ligaments and tendons in the legs) in conjunction with the check apparatus of the forelimbs and reciprocal apparatus of the hindlimbs enable a horse to relax while standing without collapsing (Adams 1966). Winchester (1943) found that standing, not recumbency, is the posture of minimal energy demand on horses. Recumbency causes come cardiac, respiratory, and other internal stress due to pressure against the substrate. Nevertheless, recumbency occurs in most horses at least once each day provided environmental conditions are not too stressful or severe.

In stabled horses, Steinhart (1937) found 11.5% of each day was spent lying down in either lateral (4.0%) or sternal (7.5%) recumbency. The stabled horses observed by Ruckebusch (1972) were recumbent 8.2% of the average 24-hour cycle. During the nighttime period alone, recumbency occurred an average of 19.9% of the time. During summer nights, Keiper and Keenan (1980) observed feral ponies spent 23.5% of the nocturnal period resting in the standing posture and 16.5% in a recumbent position. The field data gathered over a three year period by Duncan (1980) show a trend for horses to spend less time in recumbency during colder months and more time resting in the standing posture. He also found adult females rest more in the standing posture and spend less time in recumbency than any other sex/age class. As the population under observation doubled over the three years, there was a general trend for all animals to spend less time in recumbency and to correspondingly increase time resting in the standing posture.

In the standing posture, a resting horse is supported usually by only three legs (Figure 7.1) with the slope of the neck lower than when attentive and alert. The muscles relax, the ears rotate laterally, and the eyelids and lips get droopy. As slow-wave sleep proceeds, the eyes tend to close and the neck often continues to relax; in the extreme, the crest of the neck may drop below horizontal with the dorsal surface of the head sometimes reaching vertical. The individual may remain sleeping in this posture for many minutes before arousing. Contrary to the opinion of some clients on rented horses, horses do not actually sleep while walking.

Figure 7.1: Resting with weight distributed among only three of the legs.

During a period of arousal or as a horse becomes drowsy, it may go down to sternal recumbency (Figure 7.2a). While in sternal recumbency the individual may fall asleep and relax the head and neck. If sleep progresses while still in sternal recumbency, the relaxed neck often causes the mouth and lips to contact the substrate (Figure 7.2b,c). Sleep may proceed in that posture or the head may extend allowing the ventral surface of the lower jaw to rest on the substrate.

It is not unusual to see horses become drowsy and fail to initially assume lateral recumbency. They doze off repeatedly, for example, while standing or in sternal recumbency sunning themselves. After several brief bouts of slow-wave sleep, they may eventually assume lateral recumbency (Figure 7.2d).

In lateral recumbency, the side of the head and neck are placed on the substrate as the body shifts completely onto one side. The legs become somewhat extended, the eyes may then close, and as sleep proceeds the facial and skeletal musculature relaxes further. Lateral recumbency can last for as long as 60 minutes, although Steinhart (1937) reported an average of 23 minutes for the stabled horses he observed.

Similar to other mammals, including man, the horse exhibits different states of wakefulness and sleep—alert wakefulness, drowsiness, slow-wave sleep, and paradoxical sleep. Drowsiness is like a transition between alert wakefulness and slow-wave sleep. Slow-wave sleep is the initial and more frequent form of sleep and can occur in a standing or recumbent position. Paradoxical sleep is a very deep sleep occurring in lateral or occasionally sternal recumbency. Although an animal is difficult to arouse during this stage of sleep, its electroencephalographic pattern and muscular activity would suggest it is almost awake, hence the name paradoxical sleep.

Figure 7.2: Resting attitudes assumed by recumbent horses.

Ruckebusch and his co-workers (1970, 1972) have studied sleep in horses using electrocorticographic (ECoG) recordings. They found alert wakefulness as well as paradoxical sleep exhibited desynchronized ECoG recordings showing low voltage, fast activity. Slow-wave sleep was characterized by synchronized high voltage, slow activity. And drowsiness showed a mixture of both low voltage, fast activity and high voltage, slow activity. During alert wakefulness, theta rhythm characterized by periodic low frequency waves could be noted.

Heart and respiratory rates decreased as subjects progressed into deeper sleep; yet, heart rate often elevated again within bouts of paradoxical sleep (Table 7.1). During paradoxical sleep, bursts of rapid eye movement (REM) and often-times movement of limbs, ears, and facial musculature occurred. Rapid heart rate (tachycardia) and increased respiration (polypnea) were common during REM bursts in the course of paradoxical sleep. Eye closure was complete in paradoxical sleep, but not necessarily in slow-wave sleep. During the sleep period, loss of muscular tone was initially gradual then commenced rapidly about mid cycle of slow-wave sleep and remained negligible during paradoxical sleep.

Table 7.1: Cardiac and Respiratory Rate in Different States of Wakefulness and Sleep

	Alert Wakefulness	Drowsiness	Slow-wave Sleep	Paradoxical Sleep
Heart rate	43.5±5.1*	41.7±2.9	39.0±2.9*	41.8±1.7
Respiratory rate	19.6±4.8*	12.5±2.8*	9.8±1.7*	10.0±2.6

$^*P \leq 0.05$

Data from Ruckebusch et al. 1970

Drowsy horses become alert to new sounds in their environment, whereas horses in deep sleep are not easily aroused. The threshold for arousal by audio-stimulation increases by approximately a factor of 10 during slow-wave sleep compared to that during drowsiness (Ruckebusch 1972).

The daily pattern of sleep in horses varies from one environmental situation to another. Some horses sleep only at night; others utilize daylight periods. Individuals seem to have their own specific sleep-wakefulness pattern, varying less than 5% from day to day; whereas variation between individuals kept under the same conditions can be 10-25% (Ruckebusch 1972). Many observers (e.g., Tyler 1969, Welsh 1975, Feist and McCullough 1976, Keiper and Keenan 1980) have noted recumbency and sleep in free-roaming horses occur during daylight as well as nocturnal hours.

Ruckebusch (1972), monitoring stallions in barn stalls, found sleep in his experimental subjects occurred only at night. The data he accumulated on three stallions by ECoG monitoring for periods of two to three consecutive 24-hour periods per week are shown in Table 7.2 and Figure 7.3. The wakeful state occupied on the average 88.8% of the 24-hour period and 71.4% of the nocturnal hours. Drowsiness, although of short duration relative to ruminants, occurred numerous times each day. Paradoxical sleep bouts averaged over five minutes and recurred several times each rest period. Tachycardia and increased breathing during paradoxical sleep occurred independent of limb movement, suggesting they were a direct result of dream-like episodes. Bouts of leg, ear, and eye movement as well as clonic contractions of the face plus vocalizations during paradoxical sleep suggest that horses experience vivid dreams.

Table 7.2: Proportion of Time Spent in Sleep versus Wakefulness*

	10-hr Night Period	24-hr Period
Total duration and percentage		
Wakefulness		
Alert wakefulness	5 hr 14 min (52.4%)	19 hr 13 min (80.8%)
Drowsiness	1 hr 54 min (19.0%)	1 hr 55 min (8.0%)
Sleep		
Slow-wave sleep	2 hr 5 min (20.8%)	2 hr 5 min (8.7%)
Paradoxical sleep	47 min (7.8%)	47 min (3.3%)
Posture		
Standing	8 hr 1 min (80.1%)	22 hr 1 min (91.8%)
Recumbent	1 hr 59 min (19.9%)	1 hr 59 min (8.2%)
Ratio (as percentage)		
Drowsiness:Total wakefulness	26.63%	9.06%
Paradoxical sleep:Total sleep	27.32%	27.32%
Mean duration and no. of periods		
Drowsiness	3 min 56 sec (29)	3 min 29 sec (33)
Paradoxical sleep	5 min 13 sec (9)	5 min 13 sec (9)

*Average values for three stallions housed in stalls.

Data from Ruckebusch 1972

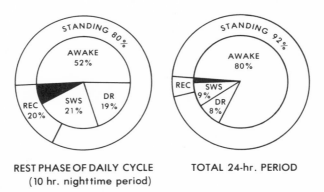

REST PHASE OF DAILY CYCLE   TOTAL 24-hr. PERIOD
(10 hr. nighttime period)

Figure 7.3: Average sleep and wakefulness pattern of three stallions moni-
tored electroencephalographically while in stalls. Outer circle shows postures,
and the inner circle represents the relative duration of sleep and wakeful
states. Paradoxical sleep is shown in black; DR=drowsiness; SWS=slow-wave
sleep; REC=recumbent. (Adapted from Ruckebusch 1972)

Diet is one of the various factors that affect patterns of sleep and wakefulness in horses. Dallaire and Ruckebusch (1974) determined that ponies housed under a controlled temperature and light regimen with free access to hay and water exhibited a total daily pattern of about four hours in sternal recumbency and one hour in lateral recumbency. When oats were substituted for hay on a ratio by weight of 1:2, the total recumbency time was increased by about 20%. Sternal not lateral recumbency accounted for the increase. Total sleep time (slow-wave sleep plus paradoxical sleep) also increased; paradoxical sleep remained about 25% of the total sleep time. Similar results occurred after two or three days of fasting with only water available.

why?

8

Ingestive Behavior

Horses are adaptable to a variety of foods and ingestion schedules. They can tolerate rather desolate conditions with a scarcity of food and water; yet, horses show a preference for grasses and grass-like plant materials as well as for a nearby water source. Unlike ruminants, their cecal digestion, high intake, and rapid food passage enable horses to adequately maintain themselves on a high fiber, low protein diet (Janis 1976). When preferred foods, such as grasses, are no longer available, their diet may include roots, browse, or aquatic plants; oftentimes a variety of foods are consumed in one day. Seasonal patterns in ingestive behavior occur in most locations.

Foals commence nursing within an hour or two of birth. Nursing declines over the next few months as time spent grazing increases. In the review that follows, nursing will be considered separately from other feeding and drinking behaviors.

FEEDING

While grazing or browsing, horses manipulate plant materials with their dexterous upper lip. The preferred plant items are isolated from adjacent materials using the upper lip, then the bundle is passed between the upper and lower incisor teeth where biting, assisted by a jerk of the head posteriorly, nips off the leaves or other parts, and with the aid of the tongue the materials are ingested into the mouth for chewing (see Figure 6.1). Other bites may be taken before a bout of chewing commences. Grains and concentrated feeds are also ingested with the combined action of tongue and lips.

Grinding of the food with the well-suited cheek teeth occurs at a rate of 1 to 1.5 times per second. The macerated material is then swallowed as one or more boluses of food passing along the esophagus to the stomach of 7-14 liter capacity. Commonly, a horse shifts its neck from side to side as it grazes slowly forward, stepping to make additional plants accessible. Selective feeding is typical.

In most cases, the neck must be lowered to enable the mouth access to the food material. The body axis is often kept parallel to the direction of wind, and a vigilance is maintained using the eyes and ears. While chewing, the neck and head are momentarily raised and observations of the surroundings are made before the grazing pattern then continues. It is not unusual for an entire herd to graze at the same time and in the same direction, maintaining an individual distance of at least one meter between each other (Figure 8.1).

Sometimes horses must dig for food by pawing with a foreleg. In winter, pawing is especially utilized in deep snow where numerous strokes may be used to clear a crater and expose plants. In a snow depth of 40-50 cm, Salter and Hudson (1979) found horses pawed an average of 9.7 times per bout (9.1 bouts per 5 minutes) compared to 5.4 strokes per bout (1.4 bouts per 5 minutes) when the snow depth was 10 cm. In shallow snow, horses push away the snow with their muzzle without the need for pawing. In arid habitats where food supplies have dwindled, horses dig up roots using pawing movements.

When feeding on submerged aquatic vegetation, a horse may need to immerse its muzzle well below the surface. Some stabled horses learn to moisten dry forage by dunking the roughage in their water supply. One mare I observed lifted hay routinely to a shelf near an automatic waterer and would proceed to dip mouthfulls of hay at the rate of 5.1±1.6 SD per minute between bouts of chewing. If the waterer was turned off, or if the roughage was fresh cut or previously moistened, the mare would not go through with the dunking routine before swallowing (Waring 1974).

Horses do not evenly graze their habitat. With each environmental setting they show preferences for certain plant species and may avoid others. Short, new growth is often favored. In pastures, a grazing pattern usually becomes evident where horses feed heavily in certain areas and utilize other portions of their available space for eliminative areas (Taylor 1954, Ödberg and Francis-Smith (1976). The areas grazed become cropped close to the ground, and taller, rough vegetation develops in the defecation areas where feeding rarely occurs. Ödberg and Francis-Smith (1977) concluded horses tend to defecate in the areas used previously and thus these areas become seldom-grazed zones. Palatability of the vegetation appears not to be as much a factor for omitting feeding in the latter areas as does the physical presence of feces. In tests made by Archer (1978), horses did not graze where feces were placed but did defecate on those sites; moreover, the horses grazed test plots treated only with urine.

The time of day as well as the total time spent feeding are dependent on the quality and quantity of food available to horses, plus such factors as exercise, lactation, weather, and insect pests (e.g., Martin-Rosset et al. 1978). Environmental disruptions, such as storms or intruders, can temporarily cause horses to discontinue grazing. Social factors also influence feeding patterns; for example, as one horse begins to graze other group members are more inclined to graze (a form of social facilitation). Stabled horses fed limited amounts of concentrates, grain, and hay generally consume their ration soon after feeding and are left without food for much of the day. With food made constantly available, Ralston et al. (1979) found ponies consumed 80% of their daily intake in an average of 10 separate meals. Each meal averaged 0.5 kg of a pelleted ration and

a.

b.

Figure 8.1: Typical feeding activity of horses showing (a) grazing and (b) foraging along surface for such items as acorns.

lasted 44 ± 10 minutes; on this diet, 38% of the 24-hour day was spent feeding. The average interval between meals was 84 minutes. Half of the intake was consumed between 0800 and 1700 hours.

Range horses typically feed long hours. Salter (1978) found feral adult horses of western Alberta fed about 75% of the daylight hours in winter and spring, whereas foals spent about 41% of their time foraging. Similarly in England during winter, ponies ranging the New Forest spent most of their daylight hours grazing and browsing; but after May, resting time increased as grazing time decreased. Then as flies became abundant in June, the ponies remained in the shade with few feeding excursions between 0900 and 1400 hours (Tyler 1969). Both Tyler (1969) and Salter (1978) noted that peak grazing activity in daylight occurred about dawn and again in the late afternoon. One or two resting periods were typical between the peak feeding times.

Grazing at night is a common activity for horses. For example, on Assateague Island along the Maryland-Virginia coast, feral ponies during summer nights were found to graze 54.6% of the nocturnal period (Keiper and Keenan 1980). Although grazing occurred periodically through the night, there was a tendency for greatest feeding activity early in the evening and again at dawn.

In Poland, Kownacki et al. (1978) seasonally sampled day- and nighttime behavior of horses on a forested reserve. They found adult horses foraged nearly 70% of each 24-hour day, with little apparent change in the total grazing time between early summer, fall, and winter. During the winter supplementary hay was utilized.

In the Camargue region of southern France, winter forage is also scarce; but in the growing season, horses feed extensively on emergent marsh vegetation. In this region Duncan (1980) found a slight tendency for mares to spend more of their time (58.5%-63.1%) foraging in all seasons than mature stallions (50.8%-59.7%).

Foals do little grazing during their first few weeks and unless utilizing a slope or hummock must spread their forelegs to allow the mouth to reach plants close to the ground. Some foals reach the plants by flexing the forelegs at the knees. As the foal develops, grazing activity increases. Tyler (1969) found a tendency for foals not only to graze more with age but also to graze significantly more ($P<0.001$) during the late afternoon (Figure 8.2). Boyd (1980) noticed a two-day-old orphan foal in its effort to survive cropped tips of grasses and brush with little selectivity.

The diets of horses vary greatly between one habitat or management situation and another. Horses also vary in their individual grazing selectivity (cf. Marinier 1980). In general, horses are opportunistic feeders, selecting the most palatable and accessible items available to them. Thus in confinement, they will accept pelleted purified diets (Stowe 1969). In marsh habitats, they will eat emergent as well as submerged aquatic plants (Ebhardt 1957, Göbel and Zeeb 1963, Tyler 1969). In woodlands, they browse on a variety of plants and will consume bark, buds, leaves, and fruits. In the fall, some ponies Tyler (1969) observed spent much of their day grubbing under oaks for fallen acorns. In other circumstances, horses may seek roots. Yet, it remains that horses, just as is assumed of their recent ancestors, are primarily grazers, chosing grasses and grass-like forages whenever feasible for the bulk of their diet.

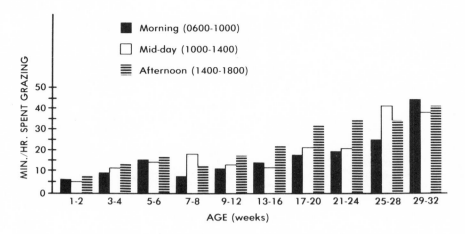

Figure 8.2: Grazing activity of foals during daylight. (Data from Tyler 1969)

The analysis of diets of free-roaming horses is typically done by microscopic techniques to determine the botanical composition of plant remains in samples of fecal material. These data are used to estimate the percentage of dry weight of the various plant species in the diets of the animals. For example, in a study of feral horses of western Colorado, Hubbard and Hansen (1976) found diets in mountain shrub areas emphasized sedges (46%) which were relatively abundant, but the diet included some grasses and a shrub (Utah serviceberry) also. In lower piñon-juniper areas, several kinds of grasses dominated the diet (Table 8.1). A single woody plant (common winterfat) was used relatively often for food, but averaged 7% or below in the diet on both plant zones. Feist and McCullough (1976) noticed horses dug up the roots of this plant.

In another study (Hansen 1976), feral horses living on the desert grassland areas of southern New Mexico showed diets emphasizing Russian thistle, grasses, and mesquite. Seasonal use varied greatly; mesquite pods and leaves made up 53% of the September diet of these horses, yet only 2% in March. In sagebrush-saltbrush-rabbitbrush areas of the Red Desert of Wyoming, grasses are mainly used for food (Olsen and Hansen 1977). Farther north in the foothills of Alberta, free-roaming horses feed on numerous plants but greatly emphasize grasses and sedges (Salter and Hudson 1979). At coastal sites, cord and beach grasses seem to be the major diet (e.g., Zervanos and Keiper 1980).

Aside from what can be considered normal diets and usual feeding behavior, some horses spend time ingesting fecal material (coprophagy), eating mud, or chewing wood. Coprophagy can be common in foals up to a month of age (Tyler 1969, Blakeslee 1974), but it wanes thereafter. Usually only one or two pellets are eaten subsequent to a bout of pawing of the material. The feces ingested are most commonly those of the mother. Such behavior may aid foals in acquiring beneficial intestinal microorganisms, yet parasites too get ingested. Coprophagy is rare in adult horses, although stallions are especially eager to investigate feces and add to existing piles. Feist and McCullough (1976) suggested that occurrences of coprophagy by older free-roaming horses may be due to food scarci-

Table 8.1: Variations in Diets of Free-Roaming Horses in Different Habitats of North America

Plant Species	Barrier Island (Md/Va)	Desert Grassland (N. Mex.)	Desert Shrub (Wyo.)	Piñon-Juniper (Colo.)	Mountain Shrub (Colo.)	Boreal Forest (Alberta)
Grasses and Grass-likes:						
Sedge (*Carex* spp.)	*		*	*	**	**
American Three-Square Sedge (*Scirpus americanus*)						
Cotton Grass (*Eriophorum viridi-carnatum*)						*
Wire Rush (*Juncus balticus*)						**
Needlegrass (*Stipa* spp.)			**	**		*
Wheatgrass (*Agropyron* spp.)		*	**	**	*	
Junegrass (*Koeleria cristata*)		**	**	*	**	
Brome (*Bromus* spp.)			*	**	**	*
Tufted Hairgrass (*Deschampsia caespitosa*)					*	*
Hairy Wildrye (*Elymus innovatus*)						**
False Melic (*Schizachne purpurascens*)						*
Indian Ricegrass (*Oryzopsis hymenoides*)			**	*	*	
Bluegrass (*Poa* spp.)			*	**	**	**
Fescue (*Festuca* spp.)					*	**
Dropseed Grass (*Sporobolus* spp.)		**				
Spangletop (*Leptochloa dubia*)		*				
Grama (*Bouteloua* spp.)		*				
Muhly (*Muhlenbergia* spp.)		*				

(continued)

Table 8.1: (continued)

Plant Species	Barrier Island (Md/Va)	Desert Grassland (N. Mex.)	Desert Shrub (Wyo.)	Piñon-Juniper (Colo.)	Mountain Shrub (Colo.)	Boreal Forest (Alberta)
Bristlegrass (*Setaria macrostachya*)		*				
Hairgrass (*Agrostis scabra*)						*
Timber Oatgrass (*Danthonia intermedia*)						*
Salt-marsh Cordgrass (*Spartina alterniflora*)	**					
Salt-meadow Cordgrass (*Spartina patens*)	**					
American Beech Grass (*Ammophila breviligulata*)	**					
Forbs, Browse, and Others:						
Rabbitbrush (*Chrysothamnus* spp.)		*				
Utah Serviceberry (*Amelanchier utahensis*)					*	
Common Winterfat (*Eurotia lanata*)			*	*	*	
Russian Thistle (*Salsola kali*)		**				
Mesquite (*Prosopis juliflora*)		**				
Saltbush (*Atriplex* spp.)		**	*			
Snowberry (*Symphoricarpos* spp.)			*			
Globemallow (*Sphaeralcea* spp.)						
Lodgepole Pine (*Pinus contorta*)						*
Horsetail (*Equisetum* spp.)						*
Moss						*

* = Seasonal or annual diet 1 to 9%.   ** = Seasonal or annual diet 10% or more.

Data from Hansen 1976, Hubbard and Hansen 1976, Olsen and Hansen 1977, Salter and Hudson 1979, Ford and Keiper 1979, and Zervanos and Keiper 1980 ties. As examples, they reported observations of mares and their offspring eating old pellets from stallion fecal piles during August and winter.

Soil ingestion, although apparently not frequent, has been observed in horses under varying circumstances. Feist (1971) recalled seeing non-feral horses ingest soil from newly plowed fields in Canada and observed a lone feral stallion on the Pryor Mountain Wild Horse Range in May eat dark gray mud from a nearly dried up puddle. Salter and Hudson (1979) found free-roaming horses in Alberta ingested throughout the year quantities of soil at salt licks established for cattle as well as at natural mineral licks. Supplementation of dietary sodium has been suggested as the primary benefit of soil ingestion (Salter and Pluth 1980).

Wood chewing, especially fence or stall chewing, can be observed in some confined horses. Softwoods as well as hardwoods are vulnerable. Destruction, of course, becomes more evident and widespread with the softer wood types. Horizontal boards as well as uprights at corners are the usual targets for chewing, wherever the animal stands for considerable time. Although some ingestion may occur, much of the chewed material falls to the ground.

Restlessness as well as dietary deficiency may justifiably be implicated in most cases of abnormal ingestive behavior. Willard et al. (1973) found ponies fed an all-concentrate diet spent more time chewing wood, eating feces, and licking salt than did ponies on a hay diet. Haenlein et al. (1966) found ponies maintained on pelleted food became nervous and exhibited wood chewing, although food was still available. Wafered food did not cause similar problems. In other studies, young horses kept on pelleted diets not only have shown wood chewing but also chewing on the manes and tails of other horses (Willard et al. 1977). To pursue the problem further, Willard et al. (1977) altered the diets and cecal pH of horses, while water and trace mineral salt were available free-choice. Compared to horses on a non-pelleted concentrated diet, horses on a mixed grass-legume hay diet spend significantly more time eating ($P<0.05$) and significantly less time ($P<0.1$) in wood chewing, coprophagy, and food searching activity. Furthermore, horses on the concentrated diet with experimentally increased cecum pH (via sodium bicarbonate infusions) spent significantly more time standing ($P<0.05$) and less time in coprophagy ($P<0.1$) than did horses fed concentrate alone. Thus, the type of diet and factors such as increased cecal acidity appear to influence abnormal feeding behavior.

DRINKING

Besides food, water is an important resource for horses. A horse ingests water by immersing its almost-closed lips below the water surface and through a sucking action pulls the water into its mouth (Figure 8.3). Swallowing proceeds at a rate of once per second. More than 4 liters may be consumed in one drinking bout. The horse then pauses to look around and may proceed to drink some more. Many sources of water are used for drinking, provided sufficient depth is available to allow for the immersion of the lips. Small pools resulting from precipitation or nearby springs suffice for many free-roaming horses. Sometimes horses create their own drinking pool by pawing a crater in sandy soil (Welsh

1973). Snow ingestion and succulent foods may help alleviate the demand for water.

Figure 8.3: Drinking response of horses showing immersion of lips below water surface. (Photo courtesy of R.R. Keiper)

The frequency of drinking varies with such factors as accessibility and physiological need. Drinking can be day or night with little fixed schedule. When water is readily within reach, horses drink small amounts several times in one day. As accessibility of water diminishes, movement to water sites becomes less frequent. Free-roaming horses which may be several kilometers from their nearest water hole generally return and drink once each day (Feist 1971) or as little as every other day (Pellegrini 1971). In extreme heat, herds may remain near water and drink more often. Przewalski horses can apparently drink as seldom as every two or three days (Bannikov 1961).

When one horse moves toward water, the activity is contagious and others tend to join in the single file procession and the subsequent activity of drinking. When space is limited at the water source, the more dominant animals drink first. As each horse finishes drinking, it tends to wait for the remainder of the social group and all move away together. Feist and McCullough (1976) found a group begins to leave after drinking 2 to 10 minutes, rarely staying longer than

30 minutes. They noticed other bands of horses waited at a distance until the group at the water hole vacated the site. Pellegrini (1971) found the bands he studied sometimes spent the night near water holes after leisurely drinking. On Assateague Island during summer, feral ponies tend to move to water just before or soon after sunset; the highest incidence of nighttime drinking occurs during the first hour of darkness, although drinking behavior is also occasionally seen at other hours (Keiper and Keenan (1980).

NURSING

The sucking reflex of newborn foals can be induced soon after birth. Tactile stimulation along the mouth or muzzle induces the tongue to protrude slightly from the lips, the head and neck extend as well as elevate, and sucking motions of the mouth begin. During the first hour following birth, the drive to suckle appears to increase progressively and the response becomes easily induced by any contact with the lips or dorsal surface of the muzzle. Foals I have kept in sternal recumbency for their first hour to permit human fondling and socialization have shown spontaneous sucking in mid air at approximately 50 minutes of age even without tactile stimulation of their muzzle.

Besides displaying the sucking reflex, neonatal foals must achieve standing and exhibit teat searching behavior before nursing can be established. Also necessary for successful nursing is the cooperation of the mare. Some mares assist the searching behavior of their foals by presenting the flank region and nuzzling the foal into position. But oftentimes mares show no assistance, and their unsteady foals may spend many minutes searching between the mare's forelegs, probing her belly and sides, or even probing surrounding inanimate objects. Foals seem to be inclined to search under surfaces at about head height while trying to maintain tactile stimulation against the top of their muzzle. Tactile stimulation, rather than visual or chemical cues, appears to be the primary basis for such searching maneuvers with the head and mouth; nevertheless, foals are probably visually attracted to large objects and finally attach to a teat because of both tactile and chemical stimulation.

As foals begin to probe near the often sensitive udders, some mares resist by moving away, or they squeal and bump the foal by lifting the stifle region of the hindleg against the foal's neck or shoulder, causing the foal to withdraw. Once a nursing routine commences, the mare's discomfort eventually wanes.

With repeated searching and head extension near the mare's flank, a foal eventually locates one or both teats and commences to briefly suckle. Of the 245 Thoroughbred foals Rossdale (1967a) studied, nursing first occurred an average of 111 minutes after birth; yet the data ranged from 35 to 420 minutes. At subsequent nursings, the movements of a foal become increasingly more coordinated and directed toward the mare's flank, udders, and teats. Nursing bouts in stabled horses tend to recur at intervals of 10 to 90 minutes.

The usual nursing posture is such that the foal places its head beneath the mare's flank while standing with its hindquarters near the mare's shoulder (Figure 8.4). Both teats can be reached without changing position. The mare often facilitates access by stepping forward yet leaving the hindleg on the foal's side in place exposing her flank. Commonly the mare turns her head and smells or licks the foal's hindquarters. Occasionally the foal's body is angled more perpendicularly to the mare's body; more rarely, some foals succeed in nursing from behind the mare by reaching between her hindlegs. If two foals nurse a mare at the same time, one foal typically assumes the common position along her side while the other reaches the remaining teat by using the between-the-hindlegs approach. An additional nursing posture which rarely occurs is for a standing foal to suckle while the mare is in lateral recumbency.

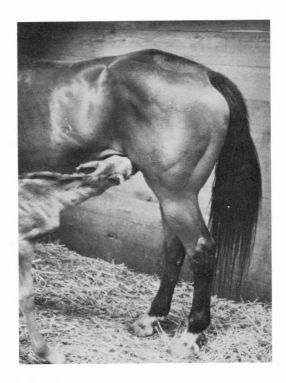

Figure 8.4: Typical nursing posture of foals.

A mare usually only allows her most recent foal to nurse. Weaning of the previous foal occurs weeks before or at the time the new foal is born. Occasionally, however, an older sibling suceeds in regular nursing subsequent to the arrival of the newborn. For example, Tyler (1969) observed a yearling nurse periodically for almost a week after the birth of its sibling, and in another case a 2-year-old filly nursed for several months subsequent to the birth of her sibling. In some instances where a mare fails to have a foal the following year, the nursing relationship is maintained with the yearling or an older juvenile. In most of these cases, weaning occurs before the next year; however, Tyler (1969) found a few immatures (2 fillies and 1 colt) still nursed as 3-year-olds.

It is rare that a mare will allow any foal except her own to nurse. We had an instance, however, at the Southern Illinois University Horse Center where a dominant mare solicited nursing by positioning her flank to a receptive 2-day-old foal of a subordinate mare after the dominant's own newborn became unable to nurse because of a mouth injury the previous day. Both of the mares accompanied by their foals had just been turned out together into an outdoor paddock from their separate foaling stalls. The subordinate mare was noticeably distressed but did not intervene while her own foal nursed the dominant mare.

Although foals sometimes approach other mares with nursing foals, most females threaten and drive the strange foals away. Sometimes a strange foal succeeds in sucking momentarily. For example, one mare Tyler (1969) observed had her own foal at her side when a strange foal approached from behind and nursed for over a minute before the mare turned her head to her own resting foal then quickly shifted her head to the other side and bit and chased the intruder. In cases where a mare's own foal dies or is removed, her next youngest offspring often fills the social void and may nurse on a regular basis. To establish a foster mother-foal relationship, horsemen have found some success by draping the skin of the mare's dead foal over the strange foal.

Young foals, after they have attained good motor coordination, solicit nursing by briskly approaching the mare while tossing the head, laying back the ears, and sometimes nickering (Tyler 1969). If the mare is not standing motionless, the foal often passes in front of the mare pushing under her neck and into the nursing position. If the mare proceeds to move, the foal may again move in front of the mother as if to gain her cooperation and quiet stance. Similar circling and frisky motions are exhibited toward recumbent mares by care-seeking foals.

Foals position themselves along either side of the mare while nursing; yet curiously in some circumstances one side is favored more than the other. My students and I have noticed that within a box stall, young foals tend to nurse more often from the same side of the mare. Some prefer the left; others, the right. For example, one foal nursed 86% of the time on the mare's right while in the stall, but it showed no such position preference when out-of-doors. Kownacki et al. (1978) noted that foals on pasture exhibited a 2:1 position preference.

Nursing during the day of birth is frequent and variable in length; but soon the bout duration becomes more constant, and frequency begins to decrease (see Figure 4.4). While nursing, foals exhibit a series of sucking bursts, then may pause, and may change teats between bursts (Francis-Smith 1978). Some pushing of the udder usually occurs. Nursing may last only seconds or for several minutes, but most observers report an average duration of 45 to 90 seconds. My data show that for the same foal, bouts are longer within the stall than out in pasture. One foal, for example, had an average nursing duration of 83 seconds in the barn compared to a 52 second average in pasture. Another foal showed a mean duration of 88 seconds in the barn verses 50 seconds in the pasture (Waring 1978).

The frequency of nursing and thus the total time spent nursing decreases with age. During summer daylight periods, Feist (1971) observed foals of feral horses nursed nearly twice each hour, whereas the yearlings that were nursing did so only half as often. Tyler (1969) observed that newborn New Forest pony foals nursed an average of four times per hour. This decreased to twice each hour at six weeks of age, once each hour at five months, and once every two hours at the age of eight months.

Nursing can occur at any hour; yet peaks of nursing activity have been noted. Schoen et al. (1976) found that a mid-morning as well as an early evening peak occurred. Nursing frequently is seen: (1) after periods of rest, (2) following some separation of mare and foal, (3) when the pair have been induced to change their location, such as being returned to their stall, and (4) after disturbances which result in the foal seeking protection and comfort of the mare.

The termination of a bout of nursing can be caused by movement of the mare; yet, often it is the foal who terminates nursing. Feist (1971) noted that in the free-roaming horses he observed, foals terminated nursing in 75% of the bouts. Tyler (1969) reported similar findings once foals were past two months of age; however, especially in the first few weeks mares terminated 30-45% of the bouts, primarily by moving away. When foals were less than a month of age, between 70% and 80% of the bouts were complicated because the mare continued to graze. Gradually this problem decreased. Biting was used by the mother to discourage nursing, especially in the fourth and fifth months, Tyler noted. Knocking and kicking were occasionally used to discourage foals from sucking.

Sucking responses of foals are not always exhibited for nourishment. Some non-nutritional attachment appears to occur on the mare when comfort itself is sought. Foals occasionally suck elsewhere than on their mother. For example, Tyler (1969) observed several foals sucking teats of unbred older sibling fillies, and once a foal sucked for over two minutes from a 5-month-old filly. Another foal was observed to suck the sheath of a gelding.

9

Eliminative Behavior

The eliminative behavior seen in horses will nearly always involve either defecation or urination. Regurgitation is virtually non-existent; the closest being the feeble discharge of incompletely swallowed food or fluids caused, for example, by a blockage of the esophagus. Behavioral alteration is usually minimal in such instances. Urination and defecation, however, do occur with specific behavior patterns and are linked to social behaviors as well. Thus the elimination of waste products is often more than a physiological discharge process; such activities frequently induce behaviors in nearby animals and tell much about the social and reproductive status of the individuals involved. When one horse eliminates, others in the social unit, especially adult males, often appear induced to also eliminate.

The amount of elimination per day as feces and urine reflects the intake of the animal in food and drink as well as factors such as ambient temperature. The daily fecal output per horse tends to be 14-23 kg (30-50 lb). Normal daily urine volume can range from 3-18 ml/kg of body weight (Siegmund 1973). An adult Thoroughbred, for example, weighing 440 kg might have an average urine output of 183 ml/hr (about 1.2 gallons per day). Of the total water consumed daily with food and drink, only about 22% will be eliminated as urine; most water loss occurs in respiration, feces, and sweat (Spector 1956).

URINATION

A horse about to urinate stops locomotion and assumes a basic posture where the neck is slightly lowered, the tail is raised, and the hindlegs are spread apart and stretched posteriorly (Figure 9.1). Foals, even neonates, attain the basic posture for urination. When positioning the hindlegs, horses keep their hindlegs in place and step forward with the forelegs. In strong wind, horses often orient upwind. Females tend to spread their hindlegs more than males, and males tend to attain more posterior stretch of the hindlegs. In both sexes, a slight squatting motion is common during urine flow. A stallion when marking other excrement with his own urine often raises his tail well above the more horizontal position occurring in typical urination. The penis is commonly extended slightly for urination.

The urination sequence lasts approximately ten seconds. After urination, the penis sometimes extends pendulant for a short time. In the female, urination concludes with a brief series of vulva contractions called winking, where the clitoris is repeatedly everted. Non-estrous mares and males return to a normal stance soon after urine flow ceases, sometimes switching the tail or shaking the body. Estrous mares tend to retain the urination stance momentarily with tail raised and the winking sequence prolonged.

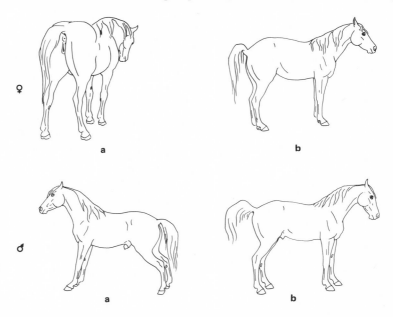

Figure 9.1: Eliminative postures of female and male: (a) urination, (b) defecation.

While in estrus, mares may urinate frequently and in small amounts (<0.5ℓ). Stallions intent on marking excrement also may urinate in small amounts as well as repeatedly. A stallion's discharge when marking is usually one or two relatively forceful squirts of urine.

In foals, urination may begin during the third hour after birth and for several weeks recurs rather often. Jeffcott (1972) found colt foals first urinated at an average of 5.97 hours (range 2.75 to 8.0), whereas filly foals urinated at 10.77 hours (range 7.25 to 15 hours). During daylight, Tyler (1969) observed foals urinated hourly for the first two weeks before the interval gradually increased, stabilizing at one year at a frequency similar to adult mares. Adult mares were found to urinate an average of once every 3.8 hours in summer and every 4.5 hours in winter (Tyler 1969). In another population of horses under somewhat similar conditions, Kownacki et al. (1978) found mares urinated an average of 7.4 times in a 24-hour period; stallions, 12.8 times; and foals, 12.5 times.

During urination, grazing ceases in the majority of instances; yet no particular site is sought. The animal in most cases seems merely to pause momentarily during other activity or as a transition occurs, such as after resting and just before grazing.

Stallions during the breeding season are often quick to investigate a mare after she urinates. Interest in the mare, however, is not prolonged if she is not in estrus. Stallions often direct their attention to the eliminated material. Feist and McCullough (1976) noted stallions responded to 50.6% of the 77 observed urinations of adult mares. The typical stallion response was to approach, smell the urine, step over it, urinate on it, and finally turn and smell again. Flehmen sometimes occurred during olfactory investigation. Although stallions in this feral population did not respond quite as often (39.5%, n = 76) to defecations of mares, the behavior of the stallions toward the material was similar. Approach, investigation, and marking were done systematically as was the case with mare urine. Urination by a stallion rather than defecation occurred in 92.1% of the responses the males made toward the excrement of adult females as well as immatures of the herd. Young males, but no females, were occasionally seen to exhibit the response shown by adult males. Boyd (1980) witnessed young females also respond to excrement in some cases and add their own urine. In New Forest ponies (Tyler 1972), where comparatively few stallions were in the population, urination by adults onto feces was rare; the adult ponies were more inclined to add their own feces. Feist and McCullough (1976) noted that urination by feral stallions was uncommon (16.8%) on the communal stallion fecal piles. Dominant stallions otherwise showed a tendency to urinate on the excrement of subordinates.

DEFECATION

The process of defecation occurs without any specific posture except that the tail is raised and often held to one side (Figure 9.1). Many times horses do not cease locomotion or continue to graze while defecating. Yet if the animal does stop, it will typically first spread the hindlges, then gradually raise its tail, protrude the anus, and finally commence elimination. The entire sequence seldom lasts more than 30 seconds, commonly less than 15. Some foals successfully defecate a few pellets before the end of the first hour after birth using a spread leg, tail up posture. The neonatal foals Tyler (1969) observed seldom defecated, yet often strained, when very young. The frequency of defecation increased with age (see Figure 4.8) and straining ceased after a few days.

After discharging feces, a horse steps forward and may switch its tail from side to side. If grazing or walking, it continues without interruption. Occasionally, however, the horse turns and smells the fecal pile. Olfactory investigation is more often seen when the horse has added fecal material to an existing pile. As with urination, stallions move to fecal piles and investigate them. Pawing of the material sometimes occurs before the animal steps over the pile and adds its own feces. A second bout of smelling concludes the marking routine.

The frequency of defecation can vary between sexes, age groups, and apparently with diet. Tyler (1969) found that during daylight hours New Forest ponies defecated on the average every 2.2 hours in summer and every 2.4 hours in winter. Kownacki et al. (1978) noticed eliminative behavior occurred somewhat uniformly throughout the 24-hour period in both sexes. In their study, defecation by stallions occurred an average of 12.8 times in a 24-hour period; in mares, it was 6.5 times; and in foals, 10.3 times in one day.

Horses in pastures not shared with other kinds of livestock commonly defecate more in certain poorly grazed areas than in areas heavily grazed. Pastures thus become partitioned into zones of short grasses as well as rough areas of tall grasses and weeds. Ödberg and Francis-Smith (1976) found the adult horses they observed spent most of their time in the short grass zones, but prior to defecation the horses would proceed to a nearby rough area, sniff the ground, defecate, and then leave the rough. Foals were less inclined to restrict defecation to the rough areas and even grazed the roughs.

Free-roaming horses tend not to limit defecation to certain areas, except for the stallions. Adult males (harem stallions as well as bachelor males) often become occupied with visits to established fecal mounds; these are thus called stud piles. Of the 186 defecations recorded by Feist and McCullough (1976), a total of 89.8% occurred at fecal piles. Sometimes younger males also use the piles. Such mounds occur periodically throughout the range and are added to by any stallion that encounters them. The largest piles are those along the routes of a number of social units, such as along a common path to water holes. Feist (1971) found the size of stud piles ranged from less than a square meter to a series of adjacent piles of successive age as large as 1.8 by 7.6 meters. The piles are often used during encounters between stallions as part of the agonistic behavior pattern.

Stallions appear to limit the amount of fecal discharge when marking fecal piles or the dung of mares, thus repeated marking can occur in a short time. Tyler (1969), for example, saw one stallion defecate on three different piles and urinate on a fourth within a 10-minute period.

10

Comfort Behavior

Comfort behavior includes such activities as sunning, scratching, rubbing, licking, rolling, and shaking. In addition, I have chosen to include mutual interactions, such as allogrooming, and the behaviors horses exhibit to minimize the effects of storms, heat, and insect pests. Many of these behaviors occur more at one time of the year than in another.

SELF-INDULGENT BEHAVIORS

Sunning

During winter months when the nights have been cold, horses on fair weather mornings seek sunny places to rest in the warmth of the direct sun rays. Each horse orients its body broadside to the sun to gain maximal exposure. Some horses stand relaxed with eyes nearly closed; others become recumbent and sometimes show signs of slow-wave sleep and even paradoxical sleep. Sunning can last for 30 minutes or more before the horse commences a different activity. Sunning is occasionally seen during other daylight hours, such as soon after a storm dissipates and warm sun rays reappear.

Shelter-Seeking

During storms, horses often seek ways to lessen the effects of the storm on themselves. They usually cease feeding and stand with their neck lowered to nearly horizontal. Recumbency rarely occurs during bad weather. In strong winds, horses stand with their hindquarters positioned into the wind, or they move to sites sheltered from the wind. Shivering sometimes occurs.

In hot weather, horses often rest in shady areas during the hottest period of the day. Sweating occurs. Prolonged periods spent in and around water may occasionally be for comfort from heat.

Wade in H_2O ?

93

Horses also seek water to reduce the effects of insects. Keiper (1979a), for example, reported feral ponies on Assateague Island moved considerable distance into the water of shallow bays or stood in the breaking waves of the surf apparently to reduce biting insects. Locomotion, such as galloping, also occurs as an anti-insect procedure. Another technique is to seek shelter of sites with low insect density. Duncan and Cowtan (1980) demonstrated that Camargue horses moved to sparsely vegetated or bare areas for resting during daylight in the summer. At such locations the attack by horseflies (tabanids) was reduced (cf. Hughes et al. 1981). The horses spent little time at such barren sites outside of the tabanid season or at night.

A degree of comfort is also achieved when horses regain the nearness of other companions after a period of separation or when a foal seeks the side of its mother when danger occurs. These behaviors are discussed further in later chapters.

Licking

Licking is occasionally exhibited by horses. It is normally used to groom in and around the mouth, but it is also a grooming procedure on accessible parts of the forelegs, shoulder, and barrel to remove primarily fluids that have soiled the pelage. Compared to many other mammals, equids make little use of licking to groom the body surface.

Nibbling

Nibbling with the incisors can vary from mild scraping of the skin with the teeth to a rapid nipping activity, where the skin is pinched repeatedly. It is a common form of grooming, apparently toward itching sensations and dried material in the pelage. Not all parts of the body are accessible to this behavior, thus nibbling can be seen primarily on the forelegs, sides, and loins (Figure 10.1a,b).

Somewhat related to nibbling is a hasty bite-like gesture toward the body surface—one of the anti-insect maneuvers of horses. The head is moved quickly toward the affected site, and the teeth may make contact with the skin. Biting motions sometimes continue at that site forming a bout of nibbling.

Scratching

The hindhoof is used by some horses to scratch the head and anterior portions of the neck. The behavior is common in foals yet rarely occurs in adults. Some mature ponies, however, retain the trait. Scratching with a hindleg is accomplished in a standing posture where the leg is raised and the hoof rubbed against the head or neck (Figure 10.1c). The neck and head are lowered and usually turned toward the raised hoof.

Rubbing

Rubbing is where the skin is massaged either by another surface of the body or against some object in the environment. The muzzle is one part of the body that is often rubbed against the forelegs or barrel apparently in response to itching sensations. Frequently, flying insects are brushed away with quick movements of a similar nature. Hindleg movement (hindleg lift) is also used toward insects and other irritants along the belly. The leg is raised swiftly, and the stifle and medial portion of the leg are rubbed briefly against the flank and belly. The raising and lowering of any leg (i.e., stamping, knocking, and hindleg lift) to scare insects off the skin can vary from slight lifts to forceful contact of the leg against the belly or substrate.

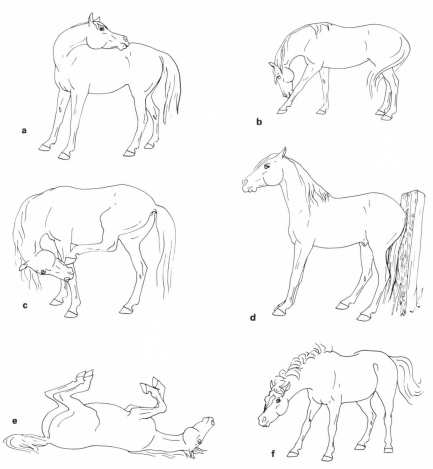

Figure 10.1: Examples of comfort behavior in horses: (a) nibbling, (b) nibbling at foreleg plus tail switching, (c) scratching, (d) rubbing, (e) rubbing back on substrate, (f) shaking.

Horses commonly use certain fixed objects in their environment for rubbing body regions such as the head, neck, base of the tail, and buttocks (Hassenberg ⤴ back 1971). Oftentimes, the horse begins with rubbing the head and neck then proceeds to rub posteriorly concluding with the buttocks and base of the tail. Fences, door frames, posts, trees, and shrubs are often utilized. The horse stands with its body touching the object and rocks back and forth rubbing a localized area of skin against the object (Figure 10.1d). Sometimes a horse will walk under low branches and let them rub the back. On other occasions, the animal may straddle a small tree or shrub and walk forward allowing the plant to rub its ventral surface. At times, horses in sternal recombency rub the region of the sternum and lower neck by rocking forward and back against the ground.

Sexual stimulation may occur in mares that rub their buttocks, tail, and vulva against objects. Tyler (1972) noted how a mare involved in such rubbing would stretch her head, sway it from side to side, and quiver her lips. Masturbation is also seen in stallions who rub their erected penis against their belly. Ejaculation sometimes occurs.

Rolling

Rolling is a specialized way of rubbing the back by utilizing the ground as the fixed object (Figure 10.1e). Rolling can accomplish other functions as well, such as dusting the pelage and for purposes related to social dominance. Many times rolling occurs at the conclusion of a period of recumbency; the animal rolls onto its back with all four legs skyward as it wiggles its back briefly against the substrate (Figure 10.2). More than one roll may occur before the horse stands. In most cases, the horse does not roll completely over but returns to sternal recumbency on the same side.

Some horses tend to roll at certain locations within their environment. The preferred sites are generally places with dry fine soil, sand, or in some cases mud. Yet rolling has been seen to occur on most substrates. Pawing of the site oftentimes precedes going down to roll.

Shaking and Skin Twitching

Shaking the whole body usually occurs after rolling, once the horse has again gotten onto its feet. Even without rolling, whole body shaking often occurs following a period of recumbency. The neck lowers to near or below horizontal, and the horse oscillates quickly the superficial musculature over much of the body vigorously shaking the pelage (Figure 10.1f) and generally releasing a cloud of dust. Shaking often begins at the head and quickly spreads posteriorly as a wave of muscular contractions. Whole body shaking occasionally occurs at other times, such as after a saddle and blanket are removed and even while being ridden.

Head shaking alone can also occur. Shaking of the head and neck happens in response to insects and other irritation around the face and ears. The direction of head shaking is basically rotational around the longitudinal axis of the body, causing strands of the mane and forelock to be flung about.

Skin twitching is the localized, rapid, oscillatory contraction of cutaneous muscles. It occurs primarily along the shoulder and forearm. Such quivering of the skin is induced by localized tactile sensations and is commonly used against biting insects attempting to alight or remain on the skin.

Figure 10.2: Photographic sequence of a mare rolling onto her back, then getting up and shaking.

d.

e.

f.

Figure 10.2: (continued)

g.

h.

i.

Figure 10.2: (continued)

cropped tail horses?

Tail Switching

Striking the body with the tail facilitates removal of insects from the hind-quarters without the need for further action with either the head or legs. Thus while flying insects are present, horses have their tail in nearly constant motion flogging the thighs, hindlegs and groin, wherever the pests attempt to land. The length of the tail governs the extent of its effectiveness. Foals and even yearlings cannot reach areas with their tail that older individuals can. The motion of the tail is primarily from side to side allowing the long strands to strike and drag over the thigh and gaskin of both hindlegs (Figure 10.1b). Occasionally the swing is forceful causing the tail to lash the barrel or to move vertically through the groin. During comfort movements, the fleshy portion of the tail is raised minimally, seldom above horizontal. The tail rarely is lashed vertically to cause the strands to strike the area of the loins and croup.

MUTUAL INTERACTIONS

Mutual Grooming

Interactive nibbling between two horses is a common form of mutual grooming (allogrooming). Licking seldom occurs between two mature horses. The two partners usually face each other, standing so that one shoulder is close to the corresponding shoulder of the partner (Figure 10.3). Nibbling of the partner may be prolonged. Feist and McCullough (1976) found the duration ranged from a few seconds to 10 minutes, but in 90% of the occasions it lasted three minutes or less. After introductory smelling, the grooming activity usually begins along the crest of the neck; it may then proceed to the withers, the shoulder, or along the back to the croup and base of the tail. Sometimes, the horses change sides.

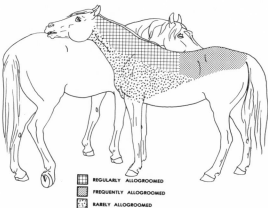

REGULARLY ALLOGROOMED
FREQUENTLY ALLOGROOMED
RARELY ALLOGROOMED

Figure 10.3: Mutual grooming.

Mutual grooming occurs primarily on areas of the body not easily reached during self-grooming activities. The dorsal portion of the neck and the withers are usual sites for nibbling. Hechler (1971) found preferences occurred in the following order for Icelandic ponies: mane (59.2%), withers (18.5%), back (9.3%), croup (5.8%), tail base or dock (4.8%), neck (1.7%), and shoulder (0.7%).

Horses tend to establish one or a few regular grooming partners. Yet, some horses never seem to allogroom (Wells and Goldschmidt-Rothschild 1979). In free-roaming herds, mutual grooming is usually only among members of the same social unit. Allogrooming can occur between females, between males and females, and even between males. It is especially frequent among immature horses. The only combination where Feist (1971) did not see mutual grooming within social bands was between herd stallions and foals. A youngster may seek such interaction, but it is rare for an adult male to reciprocate. Tyler (1969) observed that between any two partners, the more dominant individual would almost always end a grooming bout; yet dominants initiated only 38% of the mutual grooming interactions.

In some cases, foals or yearlings approach their mother while she is involved in mutual grooming and proceed to nibble at her. The mare may ignore her off-spring or commence nibbling of the youngster instead of the previous partner (Tyler 1969). Mares commonly lick their newborn foals for up to 30 minutes after parturition, but seldom does licking occur thereafter.

Several days after parturition, the mare and foal begin to allogroom. Pre-viously, the foal's nibbling and chewing on the mare was not reciprocated. The earliest age for mutual grooming Tyler (1969) observed was between a 6-day-old and its mother. Blakeslee (1974) observed a 3-day-old and a yearling in a brief allogrooming bout. By a month of age, foals begin to spend long periods mutual grooming with other foals. This trend seems to increase over the next few months.

The frequency of mutual grooming among group members shows daily and seasonal variation. Keiper and Keenan (1980) noted mutual grooming decreased significantly between 2300 and 0400 hours on summer nights, corresponding to a period when recumbency increased. During the months of spring in southern France, May is the peak month for allogrooming among Camargue mares, stal-lions, and yearlings (Wells and Goldschmidt-Rothschild 1979). In England's New Forest, mutual grooming peaks in April and again in July; it is least fre-quent in September (Tyler 1969). April corresponds to the shedding of the win-ter coat, and in July the ponies congregate in the shade.

When grouped, the tail switching of horses serves a mutual function to fend off insects. Even pairs of horses occasionally stand side by side facing opposite directions while mutually switching the forequarters of their partners. Close body contact facilitates the tail action and reduces body surface exposure. Dun-can and Vigne (1979) found horses have significantly ($P<0.01$) fewer biting horseflies on them when they were in large groups and as a consequence fewer bites. Keiper (1979a) noticed an additional strategy; ponies while clustered and facing inward took turns circling the other horses using their tails and bodies to brush away insects.

Symbiotic Relationship With Birds

Intermittent symbiotic relationships between birds and large animals, such as ungulates, are well known. In such relationships, the birds either seek ectoparasites or obtain insects disturbed by as well as attracted to the large animals. By eating ticks and biting insects or by scaring away pests, the birds benefit their symbiotic partners. With horses, such a symbiotic relationship occurs primarily with the cattle egret (Figure 10.4). During horse-egret interactions, the birds feed most often while on the ground; yet, sometimes they feed while perched on a horse's back. Horses allow the activities of the birds and are not aggressive toward them. The overt passiveness of horses to the physical contact and intimate activities of the birds is evidence that the horses may receive some comfort from the relationship and control their agonistic responses accordingly. Similar passiveness is seen in situations where a horse recognizes a person is swatting and killing horseflies that have just landed or are already biting the horse; yet when such aid is not needed, the same horse may withdraw from human handling.

mutualistic

are these cattle egret

Figure 10.4: In some regions, a mutualistic relationship develops between horses and cattle egrets. The birds benefit by feeding on insects disturbed by or attracted to horses, and the horses benefit by the removal of horseflies and other pests on their body. (Photo courtesy of Philip Malkas)

Keiper (1976b) in a study of the relationship between cattle egrets and feral ponies found the greatest feeding activity of the birds occurred when the egrets were on the ground within one meter of the ponies. Feeding strikes were not just directed at insects on vegetation. These birds directed 29.1% of their feeding strikes at insects on the ventral regions of the ponies, as follows: foreleg 10.5%, hindleg 10.2%, and underside of the body 8.4%. As many as seven egrets were seen feeding simultaneously on and near a single pony. The birds were found to be associated with stallions, mares, as well as foals in all kinds of weather and were seen to take tabanid horseflies. Not only grazing ponies attracted feeding egrets, but also ponies walking, standing, lying down, as well as nursing.

On some occasions, Keiper noted, the egrets joined the ponies while they were on sandy beaches far from vegetation, apparently to feed on insects associated with the bodies of the ponies. Egrets were often seen perched on the backs of ponies, staying up to 50 minutes at a time. Resting and preening were the common activities while perched. When egrets did feed while perched on top of ponies, feeding strikes were directed around the head of the pony (37.9%), the sides (35.1%), the back (18.9%), and the forelegs (8.1%).

Part IV

Reproductive Behavior

11

Sexual Behavior of Stallions

The stallion exhibits a variety of sexual behaviors, including being allured by a mare, testing the mare's receptivity, erection, mounting, intromission, pelvic thrusting, and ejaculation. When the animals are at pasture or free-ranging, the latter responses are seen less often than the behaviors leading up to mounting (e.g., Zeeb 1958). Males at liberty are attracted to mature mares and often seem to search for receptive individuals; they test the mares encountered for olfactory, tactile, visual, and auditory cues of sexual readiness (e.g., Tschanz 1980). Only occasional mares are receptive. The response of mares under such conditions is more often one of rejection. Only while in standing estrus will an unrestrained mare facilitate mounting and permit intromission. When not receptive, the mare will kick or show other signs of aggression as well as try to withdraw from the stallion. Most stallions thus determine the mare's receptivity with caution and become further aroused subsequent to positive feedback from the mare.

Under intensive management, the situation is usually different; a mare is commonly bred while in restraining harness, and the stallion is led to her. Prompt arousal and intromission are facilitated and encouraged. Stallions successfully achieving intromission and ejaculation under such circumstances become conditioned and soon exhibit sexual arousal, including penile erection, even before reaching the mare. Because the mare is usually restrained, testing activities by the stallion become greatly reduced, and the experienced male soon mounts. A stallion experienced at mounting a dummy (phantom) for semen collection generally becomes aroused by certain environmental factors, such as the approach of the artificial vagina or the presence of the dummy itself. Inhibiting factors can also develop and influence a stallion's behavior, causing impotence in one situation and not in other situations.

PATTERNS OF STALLION BEHAVIOR

Pre-copulatory sequences usually begin with the stallion being attracted to a mare. One of the factors involved in attraction is the urination-like stance of a mare. The stallion may emit a brief whinny and begin a prancing approach holding his head in a collected manner with neck elevated and arched. The usual gait is a trot, with legs often flexed and extended vigorously. When the mare is among other horses, the stallion may lower his neck and extend his head aggressively to drive and separate the individuals, swinging his neck from side to side and threatening to bite.

As a stallion nears a potentially receptive mare, he often emits pulsated, gutteral nickers. Upon reaching the mare, he begins to smell her head, flank, genital area, and groin; oftentimes he nips the mare's thigh or buttock. Occasionally, flehmen follows a bout of genital smelling. Depending on the cues emitted by the mare, the stallion's interest may or may not continue. When the stallion does continue, nibbling and licking of the mare's croup, hindlegs, neck, and forelegs are pre-copulatory activities which may occur as the stallion further tests the mare's receptivity and while the penis becomes fully erect (Figure 11.1).

Figure 11.1: Sexual behavior sequence of stallion testing the receptivity of an estrous mare and becoming aroused. Mounting and copulation then ensue.

Erection of the penis may begin as the stallion approaches a mare. The stallion has a vascular-muscular penis with no baculum or sigmoid flexure. Successful intromission and insemination are dependent upon sufficient sexual excitement and complete erection. Foreplay and time for full arousal are necessary for successful horse breeding. Gradually increasing tumescence of the erectile vascular tissue of the penis causes protrusion from the sheath and erection. At first, the penis protrudes so that half its length remains covered by the internal folding of the prepuce, called the sheath, which becomes smoother as erection progresses (Sisson and Grossman 1953). In mature stallions, the free part of the penis when fully erect is 30-50 cm long on the dorsal side. Full arousal of the stallion develops during the process of testing and tending a receptive mare. Repeated or prolonged flehmen, unsuccessful mounts, and withdrawal movements by the mare are among the factors that can impede or result in reduced stallion arousal and subsequent loss of erection.

Sexual arousal manifested by erection of the penis occurs also at times when an estrous mare is not being tended. Colts as young as 2 to 3 months occasionally exhibit full erection when resting or interacting with other foals (Tyler 1972). Stallions while resting in a standing posture also show occasional erection and may repeatedly swing the penis forward against the belly, for example, by rapidly lowering the croup. In some cases, such masturbation results in ejaculation. Mutual grooming, such as a stallion or colt with an anestrous mare, seems often to lead to an erection (Tyler 1972).

In sexual encounters where the male has proceeded through some pre-copulatory behavior to test the mare's receptivity and has achieved erection, mounting is commonly attempted (Figure 11.1). The stallion normally mounts from behind the mare using a sudden rearing motion with forward shuffling of the hindlegs, places his forelegs along either side of the mare embracing her flanks or sides, and rests his sternum against her croup or back. Usually the neck is lowered allowing his mouth to rest against the mare's crest or alongside her neck. Biting of these areas sometimes occurs. Young males sometimes mount from in front of the mare or along the side and may proceed to then move their hindlegs behind those of the mare.

Pelvic thrusting and intromission may or may not occur during mounting. Tyler (1972) found that when free-ranging pony stallions mounted receptive mares, intromission was achieved during the first mount in 55% of the copulations; if initially unsuccessful and the mare remained stationary, intromission was usually achieved upon the second mounting attempt. Some stallions initiate mounting before becoming sufficiently erect and thus more than one mount is required to achieve intromission. Young or impotent stallions after mounting may show little or no thrusting. Repeated thrusting is an important behavioral response required of the stallion to enable the penis to locate and penetrate the vaginal opening, achieve intromission, and subsequently facilitate ejaculation. Erection of the glans penis becomes maximal once intromission is achieved, and several pelvic oscillations usually occur before ejaculation. Asa et al. (1979) found an average of seven thrusts occurred prior to ejaculation.

Ejaculation of semen usually begins 9-16 seconds after achieving intromission. Pelvic oscillations cease just prior to ejaculation, and the concave, basin-shaped glans penis is held tightly against the end of the vagina and the cervix. Semen is forcefully ejaculated directly into the uterus (Walton 1960, Waring et al. 1975). The cessation of thrusting is an external sign of the start of ejaculation. The tail then commonly begins a series of up-down, flexing motions (tail flagging) in synchrony with and apparently induced by rhythmical shrinking of the urethral musculature. The ejaculate consists of 6 to 9 spurts or jets result-ing from contractions of the urethra (Kosiniak 1975). The respiration rate is high. Immediately following ejaculation the stallion's body relaxes, and the head droops beside the mare's neck. About 30 seconds after copulation first be-gins, most stallions have achieved ejaculation and begin to dismount. Tyler (1972) observed copulation times of 12 to 26 seconds in New Forest ponies. Pickett et al. (1970), using young Quarter Horse and Thoroughbred stallions, recorded copulation durations of 14 to 43 seconds ($\bar{x} = 27.9 \pm 7.7$ SD). As the stallion dis-mounts, the penis has begun to become flaccid and withdraws easily from the mare. Within a minute, the penis has retracted into the sheath.

A generalized summary of the response times of various stallion sexual be-haviors is provided in Table 11.1. Age, experience, season, and probably genetic background cause variation in such data.

Table 11.1: Sexual Responses of Stallions

Behavioral Response	Young (mean)	Adult (mean)
Latency of erection[1] (sec)	163	119
Latency of mounting[2] (sec)	206	101
Latency of copulation[3] (sec)	415	211
Latency of ejaculation[4] (sec)		
Natural mating	11	13
Artificial vagina	–	16
Copulation duration[5] (sec)		
Natural mating	–	15
Artificial vagina	28	–
Latency of dismounting[6] (sec)	–	8
Number of mounts per ejaculation		
Natural mating	5.7	1.4
Artificial vagina	–	2.2
Maximum number of ejaculates		
In 24 hours	–	11
In 2.5 hours	–	9

[1] Interval between stallion first seeing mare and full erection.
[2] Interval between stallion first seeing mare and first mounting.
[3] Interval between stallion first seeing mare and intromission.
[4] Interval from intromission to first emission of semen.
[5] Interval from intromission to withdrawal.
[6] Interval from ejaculation to start of dismount.

Data from Wierzbowski (1958, 1959), Nishikawa (1959), Bielanski (1960), Tyler (1972) and Pickett et al. (1970, 1976); adapted from Waring et al. (1975).

After dismounting, the stallion stands quietly behind the mare. He may yawn, stand relaxed, or begin to graze. He typically smells the genital region of the mare or the ground below. Flehmen may occur (Feist 1971, Tyler 1972). After several seconds, the pair then begins to separate. Tyler (1972) found the mare first moved away in 60% of the cases and the stallion, in 26%. Simultaneous movement occurred in 14% of the separations. On a few occasions, the mare followed the stallion when he moved away.

INTENSITY OF SEXUAL BEHAVIOR

Libido occurs throughout the year in stallions; nevertheless, the sex drive as evidenced by reaction time is higher in the spring than in autumn or winter. The intensity of stallion sexual behavior thus coincides with the breeding season of mares. If permitted, a stallion may copulate several times in one day; yet, sexual satiation does occur. Bielanski and Wierzbowski (1962) found stallions lost further sexual drive for the remainder of a day after 1-10 ejaculations; the average for their study was 2.9 ejaculates to reach satiation. A free-ranging stallion observed by Tyler (1972) attempted copulation ten times in one day, i.e., two attempts on each of five estrous mares. Six of the copulations were successful with three of the mares. No success was achieved with mounts of two other mares. In addition the stallion ignored the frequent solicitation of two 3-year-old mares during the same six hour period (1020 to 1615 hours in late April). Although adult stallions may retrieve young mares in estrus, sexual interest toward such mares typically is low.

The latency of mounting, number of mounts per ejaculation, and seminal characteristics of stallions vary seasonally. Pickett and his co-workers (1976) collected each week for 13 months two ejaculates at a one-hour interval from each of five Quarter Horse stallions. The number of mounts required for the first ejaculate did not differ from the number of mounts required for the second ejaculate collected an hour later. Nevertheless, during fall and winter months the number of mounts required for either ejaculate increased significantly compared to the spring and summer months (Figure 11.2). The reaction time from first visual contact with the mare until copulation started also markedly increased in the fall/winter period. In an earlier study (Pickett et al. 1970), the investigators found the copulation time, from entry into the artificial vagina until the start of the dismount, did not change seasonally.

Age and sexual experience affect male sexual behavior. Colts in their first few weeks of life begin to show mounting attempts. Full erection of the penis occurs in the first month and can be common by the third month. Rarely do young colts exhibit erection or pelvic thrusting while mounting. An exception was observed by Tyler (1972) where a 2-year-old estrous filly became the center of attention of a 3-month-old colt. The colt sniffed and nibbled the filly. When she spread her hindlegs and raised her tail, the colt mounted with erect penis and made numerous pelvic thrusts before dismounting. The pattern recurred repeatedly for an hour. Because of the colt's small size, intromission was not achieved.

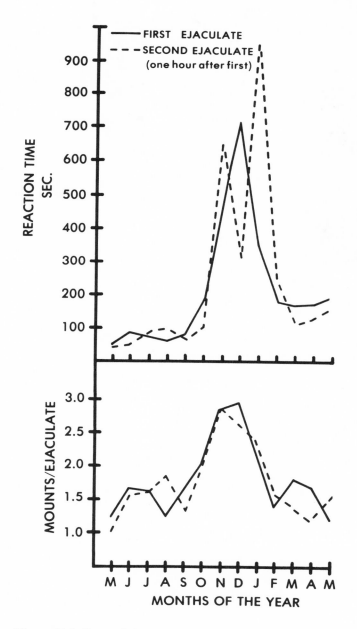

Figure 11.2: Seasonal change in reaction time and number of mounts required to collect first and second ejaculates from stallions in Colorado. Reaction time was measured as the interval from first visual contact with the mare until the start of copulation. (After Pickett et al. 1976)

As colts develop, they begin to show great interest in estrous mares. They sniff, nibble, and attempt to mount. Although young inexperienced mares are often tolerant, adult mares rarely allow colts to mount. The aggressive behavior of stallions and adult mares normally prevents young males from attempting copulatory behaviors, except with mares younger than 4 years. Harem stallions usually do not show sexual interest in such young mares of their band even when the young mares solicit during estrus.

Full growth and maturation of the male reproductive system requires several years. Spermatozoa (sperm cells) first begin to appear in the testes at 12-16 months of age (Warnick 1965). Prior to 24 months of age, most colts have low fertility. Fertility improves over the next 2 to 5 years. Thus even without social and environmental inhibitions, the reproductive role of young stallions is not likely to begin before their third summer. To become a harem stallion requires addtional years of social and physical development.

Young stallions, although possessing a high libido, lack breeding efficiency compared to experienced stallions. Wierzbowski (1959) determined the latency of erection and of mounting of young and adult stallions. The adult stallions required less time to achieve full erection after first seeing a mare. Furthermore, the average interval between first seeing the mare and first mounting was much shorter for the more experienced adult stallions, 101 seconds compared to 206 seconds for the younger group. Not only are young stallions slower to mount, they also tend to require more mounts to achieve ejaculation.

With advancing age, stallions often retain their sex drive yet have declining fertility. Under management conditions, some stallions over 20 years of age are successfully used for breeding. Nevertheless, infertility becomes more prevalent in stallions over the age of ten. Libido is not a useful measure of fertility. In some cases, fertile stallions show little sex drive because of one or more traumatic experiences associated with mating.

Stallions experienced in breeding may retain interest in mares and exhibit sexual behavior subsequent to castration. Nishikawa (1954) noted castrated adult stallions maintained normal sexual desire for 516 days after surgery. Early gelding, however, yields a male with greatly reduced intensity of normal male sexual behavior. Contrary to popular belief, there is no evidence to suggest that sexual behavior following castration is related to the presence of epididymal tissue (Crowe et al. 1977). Geldings displaying sexual interest in mares and aggressive tendencies can have testosterone and estrogen levels equal to geldings not showing such studdish behavior, according to Voith (1979b). Incomplete castration sometimes may be involved in cases where sexual behavior is retained. For example, castration of cryptorchid males can be difficult and occasionally only the tail of the epididymis is surgically removed without removing testicular tissue (Trotter and Aanes 1981). Testosterone production thus remains high. Subsequent removal of a retained testis usually diminishes sexual behavior.

STIMULI AFFECTING STALLION SEXUAL BEHAVIOR

Visual stimuli initially attract stallions to mares. The urination-like posture of a mare with hindlegs spread and tail raised seems especially effective. Frequent urination is characteristic of estrous mares as is the prolongation of the stance while showing vulval winking. During winking the clitoris is repeatedly everted on a rhythm of approximately once per second exposing light-colored membranes normally covered by the dark labial tissue. Beyond a distance of a few meters, winking appears as a bright spot flashing against a dark background.

Mares do not appear to utter any sounds to attract stallions. The splashing sounds occurring during urination may, however, attract a stallion's attention.

Some stallions reject certain mares and are sexually attracted to others. Coat color is one of the factors which may be involved. Feist (1971) noted two feral stallions in the same vicinity each sought and had in his band only mares of buckskin coloration, while a third harem band had only sorrel and bay mares. Age of the mare is another factor. Harem stallions are more attracted to full-grown mares than younger mares. When a stallion is presented with more than one mare in estrus, he tends to select the dominant mare for copulation (Asa et al. 1979).

Once attracted, the stallion seeks additional signs to test whether the mare is receptive. Visual stimuli, such as the mare's actions, continue to be involved. Estrous mares are more passive and raise their tails; non-estrous mares will usually threaten using laid back ears, bite or kick threats, as well as squealing sounds. Olfactory investigation of the mare's urine and genital region suggest odor cues further influence the stallion's sexual behavior. Young stallions usually respond little to phantoms; yet when a dummy is sprinkled with urine from an estrous mare, erection and mounting responses increase (Wierzbowski 1959). Tactile contact with the mare by nuzzling and nudging her with his head and forequarters, concurrent with positive feedback from visual and olfactory investigations, arouse the stallion so that erection develops. Inhibitory stimuli such as visual and acoustical threats by the mare, harsh tactile stimuli, and perhaps certain odors impede the progression of male sexual behavior; arousal then wanes.

In the stallion, the modification of sexual responses to visual and other stimuli can occur through learning. Stallions experienced with semen collection routines often become aroused by environmental factors other than just the mare or mounting dummy. Stallions having experienced unpleasant events when mounting or copulating may react negatively to subsequent breeding situations when a similar environmental context occurs. Among the factors involved in facilitating or inhibiting stallion performance are the site of breeding, the behavior of the mare, her size and color, dummy size and shape, handlers and handling procedures, nearby animals, and the apparatus utilized. Often the alteration of a stallion's response is situation specific. In a new environment, the response may be quite different.

The complex interaction of facilitating and inhibiting stimuli in the sexual behavior of stallions was demonstrated by Wierzbowski (1959). Stallions were tested for their reaction to a dummy and a cow as a mounting partner. Unless a phantom was treated with estrous mare urine, young naive stallions showed no response to a dummy; although when blindfolded, some of the young stallions (9%) showed a mounting response. Sexually-experienced stallions without cen-

routine breeding condition ?

sory impairment exhibited a rather high response rate (79%) to a dummy; when blindfolded, sexually-experienced stallions responded much less (38%). If presented with a cow when both visual and olfactory stimuli were impaired (using a blindfold and a nose mask containing trichloroethylene), erection and mounting responses both occurred in the three stallions tested. When other tests were conducted with only a blindfold, some sexual interest occurred upon contacting the cow. Yet when the stallions had no sensory impairment or when only the nose mask was applied, the stallions showed no sexual interest in the cow.

After intromission is achieved, tactile receptors along the surface of the penis respond to pressure of the vagina as well as to tactile stimulation caused by thrusting. Erection becomes maximal and ejaculation begins after several seconds.

Erection is, of course, necessary for successful intromission; nevertheless, erection is not essential for ejaculation. Pickett and his co-workers (1977) collected normal semen quantity and quality from impotent stallions using artificial vaginas. The stallions eagerly mounted mares and exhibited normal pelvic thrusts, copulation time, and ejaculation; yet, in these cases, no erection occurred.

Compared to cattle, the ejaculatory response of the stallion is not as sensitive to temperature. However for prolonging the life of the spermatozoa as well as stallion comfort, semen collection procedures recommend an initial artificial vagina temperature of 44°-48°C (Pickett 1974). Semen collection is most readily done using an artificial vagina while the stallion mounts a dummy or tease mare. Electro-ejaculation is of little success with stallions.

ABNORMAL SEXUAL BEHAVIOR OF STALLIONS

Abnormalities in the sexual behavior of stallions range from excessive biting and aggressiveness to the various forms of impotence and even fear of mares. Sometimes pathological causes or even congenital malfunctions are involved; yet in many cases, psychogenic factors are responsible for or contribute to aberrant sexual function and behavior. A stallion can acquire a severe inhibition toward sexual activity because of injury or psychological trauma during previous situations. Thus in cases where management procedures, pain, or unpleasant experiences have contributed to sexual abnormalities, some degree of reversibility of the acquired trait is often possible. Most patients respond well to retraining, and recoveries usually require no treatment with drugs (Pickett et al. 1977).

Impotence in stallions, that is the inability to successfully achieve copulation and ejaculation, can be exhibited in the following ways (Pickett et al. 1977): (1) Failure to obtain or maintain an erection, (2) Incomplete intromission, (3) Lack of pelvic thrusts after intromission, (4) Dismounting at the onset of ejaculation, (5) Failure to ejaculate in spite of repeated intromissions with complete and prolonged erection, and (6) Stallions that become aroused and ejaculate normally yet, in spite of high libido, cannot ejaculate normally again without prolonged sexual rest.

Not all cases of impotence can be detected from afar. For example, intromission can occur and using behavioral indicators it may appear as if ejaculation occurs; yet, palpation of the urethra during copulation may be necessary to detect that ejaculation is not occurring for a stallion having a history of poor reproductive success. The fertility of such a stallion is normally questioned long before consideration is given to the possibility of impotence. Thus proper diagnosis is important.

The diagnosis and treatment of a stallion for impotence may require a variety of tests and procedures. Libido, erection, intromission, thrusting, and ejaculation must each be evaluated. Semen collection using an artificial vagina and subsequent seminal evaluation is advisable to determine semen quantity and quality. The stallion's sex drive and behavioral patterns should be checked with several different mares in estrus. The stallion may show a preference toward mares of only a certain size or color or show interest only within a limited time of year. During examinations, the rapidity and extent of erection and other sexual responses can be evaluated. The presence of other stallions, additional horses, or particular handlers may facilitate or inhibit the stallion's responses and should be checked. Some stallions are inhibited in one situation but not in another. Stallions showing fear or extreme aggressiveness may require evaluation using a phantom. Impotent stallions should also be observed for evidence of masturbation, and whether it is a factor in the stallion's reproductive problem. In general, a thorough review is needed of the stallion's health, handling, and breeding background.

Pickett et al. (1977) recommended the following questions when diagnosing abnormal sexual behavior:

a) What is the stallion's breeding history?

b) What type of breeding program is used (i.e., pasture, hand-mating, or artificial insemination)?

c) Does the stallion's sex drive appear normal?

d) What was the stallion's sexual behavior prior to the onset of the abnormal pattern?

e) Has the stallion been observed masturbating? If so, what treatment has been initiated in an effort to control the problem?

f) How many mounts are generally required per ejaculation?

g) Does the stallion dismount at the onset of ejaculation?

h) Does he show preference for certain mares, such as mares of a particular color, stage of estrus, and so on?

i) Does the stallion object to breeding certain mares?

j) Has he been injured or frightened during teasing, mating, or while exhibiting aggressive sexual behavior?

k) Has any scrotal swelling been observed?

l) Has the stallion experienced pain or discomfort during mounting, copulation, ejaculation, or upon dismounting?

m) Has the stallion ever had laminitis or other forms of lameness causing difficulty in mounting?

n) Has he had surgery, recent illness, or exhibited any kind of unusual behavior patterns?

o) What drugs have been administered?

p) How frequently was the stallion used for breeding as a 2 or 3 year old, and how frequently has he been used this season or the season the problem was first noted?

q) What methods of discipline are used, and when are they used?

Inhibition of sexual behavior may require much work to overcome; yet the alteration of certain environment conditions, such as a change in handlers, breeding site, or methodology, may achieve success without the need for reschooling. For example, Veeckman (1979) reported the refusal of a stallion to mate mares during the first postpartum estrus (foal heat) was overcome by rubbing fresh feces of the stallion on each mare. During subsequent estruses, the mares were covered by the stallion with few difficulties. Sometimes the faulty application of an artificial vagina inhibits a stallion, whereas another technician has no difficulty collecting semen from the same stallion.

The problem of failing to obtain or maintain an erection can be more than a situation of the lack of libido. Stallions with very high libido as well as those with little sex drive may have this problem. The experiencing of severe genital injury or other trauma during sexual behavior is often involved in this syndrome. Pickett et al. (1977) described some examples they examined. One stallion had reportedly experienced penile paralysis three years earlier following the administration of a tranquilizer. When tested, the stallion in the presence of an estrous mare showed interest in mounting and, once mounted, achieved ejaculation; yet the penis was not erect and protruded only 15-20 cm. Another stallion who had been kicked on the penis by a mare during mating also would not attain an erection. The stallion would tease effectively as well as mount and thrust normally. However, in this case, massage therapy was initially needed to induce ejaculation; yet the penis remained flaccid.

In stallions maintained for breeding, masturbation is usually considered and treated as an abnormal behavior. Frequent masturbation may reduce sexual drive and cause the stallion to refuse to mount mares. Stallions that are adjusted to the breeding procedure will not normally masturbate with sufficient frequency to reduce libido (Pickett 1974). A stallion ring is commonly used to prevent masturbation. It is usually effective when properly fitted and must be removed before breeding. The ring, often made of plastic, is fitted over the end of the flaccid penis about three centimeters above the glans. When properly applied, urination is not inhibited, but the constrictive force the ring causes upon penile enlargement effectively discourages erection. Occasionally other devices, such as restrictive cages applied around the glans or wire brushes harnessed below the belly, have been used to irritate the penis when the stallion tries to achieve full erection or to masturbate.

Incomplete intromission or lack of pelvic thrusts after intromission can be caused by poor libido as well as by prior injuries or pain associated with breeding. Discomfort associated with laminitis or a persistent injury can sometimes be involved. Overuse of stallions at 2 and 3 years of age seems also to result in this problem (Pickett et al. 1977). A hereditary basis for incomplete intromission and lack of thrusting has been suggested by Hafez et al. (1962).

Dismounting at the onset of ejaculation seems associated with stallions that have good libido but a past experience of injury during copulation. A stallion treated by Pickett et al. (1977) had suffered a severe shoulder injury inflicted by a kick from another mare while the stallion was mating. Although subsequently aroused while a teasing bar or fence separated him from a mare in estrus, the stallion would show indifference to mares when in direct contact without such protection. After his confidence was partially restored around a mare, the stallion proceeded with sexual behavior yet would dismount at the onset of ejaculation. Several more weeks of additional retraining with a mare were required to return the stallion's behavior to normal.

Failure to ejaculate in spite of complete erection and energetic intromissions is occasionally reported and may have an organic or psychogenic basis (Figure 11.3). In some cases reported by Pickett et al. (1977), prior injury during copulation appeared to be a major factor inhibiting the stallion's response. Other cases, however, seem to have as a basis a previous spinal injury or some inhibition caused by handlers or tack during hand breeding. Rasbech (1975) treated six stallions with drugs when alteration of environmental conditions, including number of mares served, failed to correct ejaculatory disturbances. Two of the stallions subsequently recovered normal sexual function; one had been treated with several doses of pilocarpine, and the other, with repeated doses of ephedrine.

Some stallions show normal sexual behavior for awhile, but then begin to not ejaculate although their sexual drive remains high. Some Belgian stallions have shown such an abnormality, especially at the peak of the breeding season (Vandeplassche 1955). Sexual rest of several days normally returns such stallions to full potential only to see a return of the problem when a frequent breeding schedule is resumed. The number of mounts required to achieve ejaculation greatly increases as the problem returns. The underlying cause is not always apparent. Recurring discomfort during copulation or inappropriate handling may be involved.

Excessive aggression is an aberrant behavior problem of some stallions. Frustration of high libido stallions caused by incomplete sexual interactions and repeated non-ejaculatory copulations appears involved in most cases. Tyler (1972) noted that in the spring, stallions recently turned out to the New Forest from winter confinement attempted to mount and copulate even non-estrous mares. The stallions became increasingly aggressive and in some cases attacked young mares and even fatally mauled foals. Stallions that copulate vigorously but fail to ejaculate often become increasingly aggressive. One older stallion reported by Pickett et al. (1977) attempted to bite, kick, and strike mares after non-ejaculatory copulations developed. The savage attitude was even shown toward a phantom. The stallion's attitude became progressively worse until on one occasion he was permitted to charge the phantom and fell. When immediately presented to the phantom again, he mounted cautiously and ejaculated into an artificial vagina. Normal behavior and ejaculation were maintained with the phantom during the next two weeks. Subsequently, the stallion was returned to the owner and used successfully to breed mares.

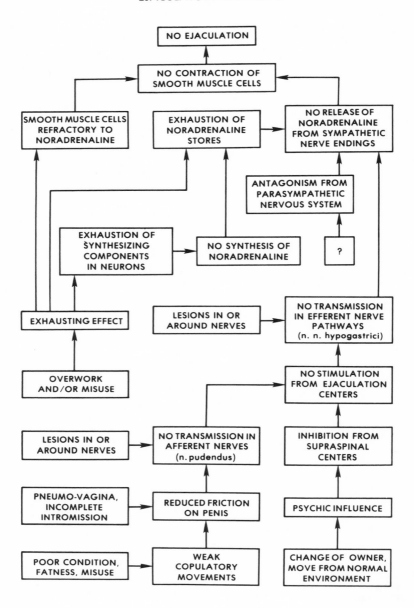

Figure 11.3: Interaction of environmental, behavioral, and physiological effects in ejaculatory disorders. (Adapted from Rasbech 1975)

During the non-breeding season, Pickett et al. (1977) observed that there is a greater tendency for certain stallions to excessively bite and strike mares prior to and during copulation. The effect of season on the number of mounts and the time needed to achieve intromission was also most pronounced for those stallions.

Besides the sometimes imprudent use of stallions for breeding or semen collection during the fall and winter, Pickett (1974) identified additional management practices as potential causes of altered sexual behavior. These are: (a) overuse of stallions of 2 to 3 years of age, (b) unduly rough handling of stallions during breeding and not permitting some aggressiveness, (c) isolation of stallions from other horses during the non-breeding season, (d) using a stallion excessively as a teaser, and (e) forcing the stallion to breed a mare when he shows considerable objection. Young stallions are especially vulnerable to the effects of early experiences in breeding. Pickett et al. (1977) recommend that a mare be hobbled and twitched to prevent kicking injuries to a stallion and that young stallions be introduced to the breeding routine gradually over a period of several weeks.

12

Sexual Behavior of Mares

A young mare's first estrus, the period of sexual solicitation and receptivity, occurs between 8 and 24 months of age. The event is used as a sign that puberty has occurred. Under management conditions with good nutrition, fillies normally reach puberty in 12 months (Ginther 1979). However, under open range conditions, many mares are in their third spring or summer before they exhibit estrus. Harem stallions tend to ignore the solicitations of these young mares. Mating with young stallions does occur, but conception is low. In Tyler's study (1972), one mare out of 107 foaled when 2 years of age (0.9%), and only 14 out of 104 of the 3-year-old mares foaled (13.5%). Breeding and foaling tend to occur at a time of year when food supply and other environmental conditions are optimum for the development and survival of the foal, including gestation and lactation success. Reflecting the seasonal breeding pattern, most foals are born during the spring after a gestation of slightly more than eleven months.

PATTERNS OF MARE BEHAVIOR

Mares are generally considered seasonally polyestrus, cyclically showing periods of diestrus (sexual quiescence) and estrus during the spring and summer then a prolonged anestrus where the reproductive physiology goes dormant in the late fall and winter. During the breeding season, the estrous cycle recurs approximately every three weeks, consisting of 5-6 days of estrus and about 15 days of diestrus. Ovulation tends to occur 24 to 48 hours before the end of estrus (Hughes et al. 1972b).

Considerable variation occurs in the cycle length and character between mares as well as seasonally within a given mare (Figure 12.1 and Table 12.1). Some mares under management conditions exhibit estrus periodically throughout the year; yet in some of these cases, ovulation is limited to the breeding season. Ginther (1979) concluded the reproductive season of pony mares is much more delineated into ovulatory and anovulatory seasons than in horse mares. During pregnancy some mares may show a bout of estrus. Once parturition has occurred, a mare may ovulate in 4 to 18 days. The mare's sexual receptivity at this time is often called foal heat and begins on the average 8 days postpartum (Matthews et al. 1967). Free-ranging mares are occasionally seen to mate within hours after foaling.

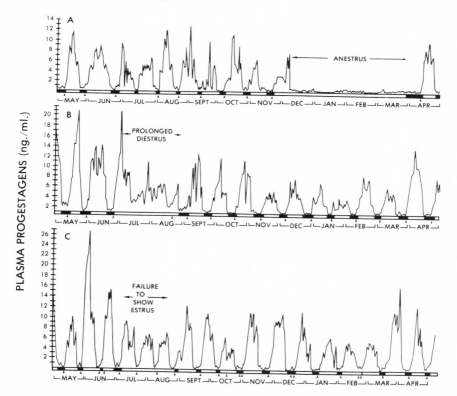

Figure 12.1: Variations in reproductive cycles of mares showing plasma progesterone levels, periods of estrus (dark bands), and ovulations (arrows): (a) Typical annual pattern with anestrus, (b) Periodic estrus throughout year with occasional prolonged diestrus, (c) Cycles with irregular ovulations and periodic failure to show estrus. (Adapted from Hughes et al. 1972a, Rossdale and Ricketts 1974, Stabenfeldt et al. 1975)

While in diestrus as well as anestrus, a mare is not receptive to the testing and sexual advances of a stallion. As the stallion approaches, her ears are laid back, she exhibits restlessness, and tail switching often occurs (Table 12.2). She avoids the stallion by moving away, or when contacted she suddenly squeals and threatens the male using bite threats as well as undirected striking and kicking. If the stallion persists with teasing, the mare no longer just threatens but directs her attack to the stallion's body.

Table 12.1: Characteristics of Mares in Breeding and Non-Breeding Seasons of the Year*

Characteristic	Mean±SD	Range
Breeding Season		
Length (days)	152±50	78–288
No. of ovulations/mare		
Total	7.2±2.0	5–10
Associated with estrus	6.8±2.4	4–10
Quiet	0.4	0–1
Double	0.1	0–1
No. of split estrous periods/mare	0.9	0–2
No. of anovulatory estrous periods/mare	0.1	0–1
Length of estrus (days)	7.1±4.2	1–26
Length of diestrus (days)	16.3±2.9	11–25
Length of interovulatory interval (days)	23.3±3.1	17–33
Anestrous Season		
Length (days)	214±50	138–288
Unseasonable estrus		
No. of periods/mare	7.1±6.6	0–19
No. of days/period	2.3±2.5	1–14
No. of days/mare	16.6±17.4	0–43

*Sample of 14 mares ranging in age from 4 to 15 years.

Data from Ginther 1974

When estrus occurs, a mare becomes relatively docile in the presence of a stallion. She allows the stallion to smell, nuzzle, and nibble her. Occasionally a mare may squeal and paw the air, only to then turn her head and touch the stallion's muzzle. Oftentimes an estrous mare shows alertness and increased activity. Urination, generally in small quantities, is frequent; Asa et al. (1979) observed one mare urinate a maximum of 21 times in one hour while in the presence of a stallion. The urination posture with hindlegs spread and tail raised tends to be prolonged after urination and is often repeated during estrus (Figure 12.2). Upon assuming the urination posture, the mare may periodically squat by lowering the pelvis. Winking, i.e., the eversion of the vulva exposing the clitoris, occurs repeatedly during the presenting stance (Figure 12.3). Fraser (1970) summarized the courtship activities of horses into four pre-coital phases: (a) greeting with nasal contact, (b) active interchange of tactile and vocal responses between stallion and mare, (c) estrous display by the mare as well as (d) her passiveness.

Associated with estrus in the mare are changes in the genital tract. For example, the vulva may become elongated, and the labia tend to swell slightly. As estrus proceeds, vaginal fluid increases and becomes less viscous. Vascularity of the lining tissues and cervix increase giving the membranes a red coloration. The cervix changes from being tightly closed to being relaxed and open at full estrus.

Table 12.2: Frequency of Behavior Patterns of Mares in Estrus and Non-Estrus While Individually Teased With a Stallion*

Behavior**	Percent Occurrence During Estrus***	Percent Occurrence During Non-Estrus
Raised tail	97.9	12.2
Urinated	53.9	7.0
Winked clitoris	87.1	10.2
Remained calm	89.0	7.4
Nuzzled stallion	12.9	5.1
Posturing	72.3	1.8
Kicked	10.8	54.5
Bit stallion	3.2	34.2
Held ears back	17.2	85.4
Switched tail	10.5	82.3
Moved about	20.1	93.4
Shook head	2.4	17.9
Pawed	2.9	27.6
Raised in front	0.9	7.8
Raised in rear	10.3	40.2
Vocal response	34.5	65.7
Snorted	0.9	13.1
Squealed	33.6	52.6
When mounted:		
Stood with tail up	100.0	1.6
Stood with tail down	—	7.2
Did not stand	—	42.6

*Based on 581 determinations during estrus and 2181 during non-estrus using 20 mares.

**Teasing technique allowed mounting by stallion to occur.

***Estrus was defined as the mare standing firmly with tail up while being mounted, plus one or more of the following: (a) winking of the clitoris during teasing, (b) urinating during teasing, or (c) tail raising before being mounted or after being dismounted.

Data from Ginther 1979

Plasma levels of the hormones estradiol, androstenedione, and luteinizing hormone peak during estrus at or shortly before ovulation (Table 12.3), whereas plasma progesterone levels fall rapidly before estrus and remain low until diestrus returns (Noden et al. 1975). The levels and interaction of at least some of these hormones are involved in estrus as well as in diestrus. Estrus can be accentuated with the administration of estradiol and suppressed by combining progesterone with estradiol or by using progesterone alone (Ginther 1979). When prolonged corpora lutea occur, estrus is inhibited but not follicular development or ovulation (Stabenfeldt et al. 1975). Since both ovariectomized mares as well as mares during the non-ovulatory season frequently exhibit estrus and copulate, and since experimental destruction of corpora lutea shortens diestrus, the accumulated evidence strongly suggests that progesterone may function, in part, to inhibit sexual behavior (Asa et al. 1980).

Figure 12.2: Mare exhibiting receptive stance and tail raising typical of estrous behavior. (Photo courtesy of P.D. Rossdale)

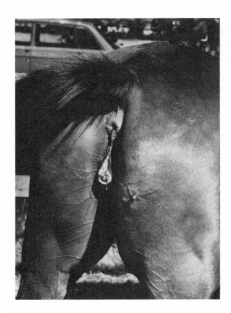

Figure 12.3: Eversion of the vulva and clitoris in an estrous mare constituting the behavior called winking. (Photo courtesy of P.D. Rossdale)

Table 12.3: Average Plasma Hormone Concentrations During the Estrous Cycle of Six Mares

Stage of Cycle	Progesterone (ng/ml)	LH (ng/ml)	Estradiol (pg/ml)	Estrone (pg/ml)	Androstene- dione (pg/ml)
Before estrus (5 days)	17.1±2.3	71±20	2.2±1.0	11.0±2.6	190±20
Before estrus (2 days)	5.3±2.0	102±11	5.6±0.6	9.2±0.9	180±30
Onset of estrus	0.7±0.2	323±78	7.1±1.6	9.7±0.8	180±30
Before ovulation (1 day)	0.4±0.1	661±113	11.5±2.5	12.5±2.3	380±70
Ovulation detected	0.8±0.5	918±199	6.8±1.7	11.2±1.7	220±30
Onset of diestrus	5.0±1.3	806±124	4.7±0.9	11.8±1.6	190±20
Mid-diestrus (7-9 days after ovulation)	13.6±2.2	123±35	4.3±0.7	10.0±1.1	210±40

± SE

After Noden et al. 1975

When a stallion does not tend an estrous mare, she may eventually move to the stallion and even assume the solicitous stance near him. If initially ignored, she grazes nearby and resumes the estrous stance periodically. Copulation typically ensues. Young mares, when in estrus, may leave their social group when ignored by the harem stallion and seek other males. Sometimes they join a bachelor male in the vicinity. When a range has only a few harem stallions, it is not unusual for an adult mare of a stallion-less group to depart temporarily from her companions during full estrus and seek contact with an adult male. Although rare, mares in estrus occasionally mount or are mounted by other mares.

Most horse breeding operations routinely tease mares using a stallion and look for positive and negative behavioral indicators of estrus. Records are commonly kept on each mare's cyclic patterns since within-mare variability is relatively low; the most likely time for ovulation can then be estimated. Teasing is a well-accepted method of estrous detection, but not all horse owners have a stallion available for teasing. In an attempt to find a reliable, efficient, and economic method to determine estrus without using a stallion, Veeckman and Ödberg (1978) studied the possible application of acoustical and tactile stimuli. Acoustical stimulation consisted of stallion courting sounds played back for a period of 2 minutes 1-2 m from the test mares. Tactile stimulation consisted of manual manipulation of the mare's neck crest, flank, or external genital region. A combination of acoustical stimulation with tactile stimulation of the flanks and external genitalia provided the best reactions. Signs of standing still, raising the tail, and spreading the hindlegs were readily elicited during estrus; during diestrus, the indicators of kicking and squealing occurred.

The urine of a mare in estrus seems to contain an odor that facilitates the interest of the stallion. The urine discharge and genital region of an estrus mare during teasing typically receive investigation by the stallion. A stallion that is then not sufficiently aroused to mount the mare may prolong the smelling of the genital region or the urine on the ground and subsequently exhibit a flehmen response. Flehmen per se does not indicate a high sexual interest. It is a means of testing the mare's condition. Further arousal may or may not follow. Although odors may serve as distant sexual attractants, observational data indicate the stallion when not approached by an estrous mare is initially attracted to her by visual cues, primarily the prolonged tail-raised, urination-like stance with vulval winking.

A bout of mutual grooming may precede copulation in some instances. Inexperienced mares, such as young mares in their first estrus, display the estrous posture near stallions but then tend to exhibit a fear response when approached by an interested stallion. They may move away or show a snapping response. Thus allogrooming, Tyler (1972) observed, seems to overcome the apprehension of the mare; copulation successfully follows.

During copulation the receptive mare generally retains her position, with legs spread to maintain her balance. Her ears are usually up, the eyes remain attentive, and the mouth is closed or opened slightly. The mare's neck maintains a moderate level, neither drooped as is characteristic of some equids nor elevated. Commonly mares turn their head slightly to observe the stallion; Asa et al. (1979) found such looking frequently occurred during ejaculation.

After the stallion dismounts, the mare often is the first to move away [e.g., in 60% of the matings Tyler (1972) observed]. Usually the mare moves only a few meters and may soon return to the proximity of the stallion and continue to show tail raising and frequent urination. The stallion, however, does not show further sexual interest in the mare for many minutes. He may investigate the ground and spilled ejaculate. Under field conditions, other behaviors and events usually then intervene and the couple thus moves apart. Grazing commonly ensues.

INTENSITY AND DURATION OF ESTRUS

The intensity of estrus varies between mares and oftentimes within-mare variation is apparent. A mare's receptivity may peak and wane more than once during estrus or remain relatively constant. Each individual mare tends to show her own characteristics which are often similar from one cycle to another. A daily rhythm of sexual receptivity is not apparent; copulations occur at any hour and usually several times during estrus. Mares are, however, especially solicitous and receptive 1-3 days before the ovum is released from the ovary. Ovulations may occur primarily at night (Studiencow 1953, Witherspoon and Talbot 1970); yet, the data are not conclusive (cf. Ginther et al. 1972). Although mature follicles can rupture at any time in the cycle, ovulation is typically associated with the end of an estrous period, 12 to 72 hours (mean = 36 h) before the end of estrus (Ginter et al. 1972). Sexual receptivity usually decreases after ovulation until estrus ceases. Wallach (1978) found evidence that sexual receptivity may begin to decrease even before ovulation.

The duration of estrus is not rigid but is commonly within a range of 5 to 15 days. Nevertheless, extremes of 1 to 50 days have been reported (Rossdale and

Ricketts 1974). Trum (1950) and Ginther et al. (1972) found the length of estrus decreases as the breeding season progresses into summer, whereas diestrus increases in length. Toward the end of the breeding season the trend reverses. The curvilinear relationship is shown in Figure 12.4. Thus in mid-summer, estrus is relatively brief especially compared to early in the season. The length of diestrus changes correspondingly so that the length of each estrous cycle and the interovulatory interval vary little throughout the breeding season.

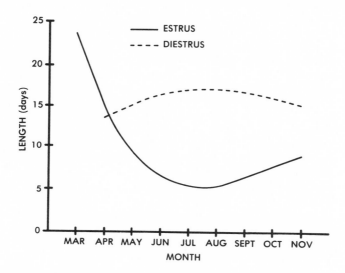

Figure 12.4: The effect of season on the length of estrus and diestrus in the northern hemisphere. (After Ginther 1974)

Veterinary procedures, such as rectal palpation, may influence the duration of estrus. Voss and Pickett (1975), for example, found estrus lasted longer in non-palpated mares than in the mares palpated.

In some mares, estrus is manifested but then ceases for a day or so before sexual receptivity again returns. Since the mare appears to be in one estrous period, such a phenomenon is called split estrus. The frequency of split estrus varies. For example in one study, Ginther et al. (1972) found split estrus occurred in 4.9% of the estrous periods, whereas with other mares (Ginther 1974) the occurrence was 12%.

Prolonged estrus can also occur lasting several weeks. Such lengthy estrous periods occur more often at the start of the breeding season. Poor nutrition or other physiological disruptions are usually considered the cause.

Mares can ovulate without showing estrus. It seems typical of some mares. They ovulate at regular intervals but fail to show clear evidence of sexual receptivity. The incidence of such covert or silent estrus is about 7%. The causes of diminished estrous behavior in mares are not known; yet, concentrations of circulatory hormones may be involved. The morphological development of follicles does not seem to be a factor (Ginther 1979).

CONTROL OF THE ESTROUS CYCLE

As with other farm animals, researchers have sought ways to influence the estrous cycles of mares. Two goals are especially sought: (1) to regulate estrus and ovulation so as to increase conception and reduce the number of matings required; and (2) to facilitate breeding outside of the normal breeding season. The first reason is to increase breeding efficiency; the second, is to adjust to the whims of mankind. In many horse breed associations, foals become one year of age on January 1 regardless of the actual birth date. Thus, to have horses that can successfully compete at the track or in the show arena as yearlings or as 2 and 3 year olds, breeders attempt to have conception occur in February or early March at a time when normal ovarian activity and ovulation may not yet be re-established.

In any attempt to control the estrous cycle, it is important to realize that inducing estrus does not assure reproductive benefit since sexual receptivity without ovulation can occur (some ovariectomized mares even show estrus). Likewise, inducing ovulation without achieving estrus is also unproductive. Ova are oftentimes shed from the ovaries of mares outside of estrus and probably are seldom fertilized. Ovulatory estrus is the goal of manipulations; so conception will result.

Intrauterine Saline Infusion

Infusing the uterus with a saline solution is a technique that has been used to induce estrus in mares. Only mares beyond the fourth day of diestrus will respond and come into ovulatory estrus. Infusions administered between 5 and 9 days after the start of diestrus hasten the onset of estrus and reduce the interovulatory interval (Arthur 1970, 1975; Ginther and Meckley 1972). Non-cyclic mares in prolonged diestrus have responded by showing ovulatory estrus 3 to 9 days after infusion; nevertheless, infusions have not been effective in shortening seasonal anestrus (Arthur 1975).

Uterine infusion in the diestrous mare induces premature regression of corpora lutea 4 to 5 days old or older (Neely et al. 1975). Thus plasma progesterone levels normally drop considerably following such treatment.

Photoperiod Manipulation

The use of artificial lighting to induce estrus during the winter non-cyclic period has met with success. For example, beginning two or more months prior to the normal breeding season, the lighted portion of a mare's photoperiod is increased in a stepwise manner or as a single major increase using 200 to 400 watt bulbs (Burkhardt 1947, Nishikawa 1959, Loy 1967). Under such treatment, mares tend to show an early onset of the breeding season. Sharp et al. (1975) exposed seven pony mares to an increasing light and temperature regimen during the winter months and found estrus became evident in all seven mares; ovulation occurred in two. Seven control mares on the same diet and housing conditions, but kept under the winter photoperiod and temperatures, did not exhibit estrus or ovulation.

Hormone Injection

A variety of hormones and similar chemicals have been used to manipulate the estrous cycle of mares. Experimentation will undoubtedly continue to be active as new compounds are identified and developed.

Human chorionic gonadotropin (HCG), chemically similar to luteinizing hormone, has for decades been used to induce ovulation in mares. It is effective provided there is a follicle sufficiently mature (>2.5 cm) to ovulate, otherwise luteinization of an immature follicle may result (Rossdale and Ricketts 1974). HCG administration early in estrus will not only induce ovulation in approximately two days, it will also shorten estrus (Loy and Hughes 1966). Intravenous injection of synthetic gonadotropin-releasing hormone (Gn-RH) also advances ovulation and shortens estrus (Irvine et al. 1975). Injections of pregnant mare serum gonadotropin (PMSG) appear to not reliably influence equine ovarian activity (Rossdale and Ricketts 1974).

Estrus can be induced by estrogen injection (e.g., diethylstibestrol); yet the effectiveness of estrogen administration alone is variable. If a mature follicle is present, an estrogen injection can induce ovulation before behavioral estrus occurs (Rossdale and Ricketts 1974). Azzie (1975) found subcutaneous implants of estradiol benzoate deposited during the normal anestrous period of mares induced estrus within 2 to 4 days; nevertheless, the mares returned to anestrus after two weeks and developed masculine characteristics until implants were removed. Their masculinization included prevalent fighting and teasing behavior when with other mares.

Estrus and ovulation can be blocked in mares using intraperitoneal injections of an antiserum containing antibodies against both follicle-stimulating hormone and luteinizing hormone when treatment occurs during estrus (Pineda and Ginther 1972). Estrus and ovulation can also be blocked by intramuscular injection of progesterone administered in daily doses of 100 mg or higher if begun mid-cycle during the luteal phase (Loy and Swan 1966). Doses of 50 mg per day were found to prevent estrus but not ovulation. Daily administration of exogenous progestogens, Loy and Swan found, can neither stop estrus nor block ovulation if treatment is begun on the first day of estrus. Progesterone administration on days 5 through 16 during a 23-day sequence of Gn-RH treatment brought acyclic mares into ovulatory estrus 24-38 days after treatment began in late winter anestrus, Hughes (1978) reported. Progesterone injections on a daily basis after parturition can be used to delay postpartum estrus and ovulation (Loy et al. 1975).

Estrus that occurs in some ovariectomized mares cannot be attributed to hormones of ovarian origin. The source of influential steriods in such cases is likely the adrenal cortex. In a study supporting this, Asa (1980) administered to ovariectomized mares a synthetic corticosteroid called dexamethasone which suppresses the synthesis of steroids in the adrenal cortex. Some experimental animals were given dexamethasone plus estradiol in case the effect of dexamethasone was other than on the adrenal gland. The incidence of estrus and thus copulatory behavior was significantly reduced in mares treated only with dexamethasone.

Prostaglandin $F_2\alpha$ ($PGF_2\alpha$) given subcutaneously induces resorption of corpora lutea in otherwise normal mares injected on day 6 of diestrus, and mares return to ovulatory estrus 3-4 days after treatment (Douglas and Ginther 1972). Hurtgen and Whitmore (1979) concluded that endometrial biopsy had similar effects on mares because of stimulating the release of prostaglandin. Allen and Rossdale (1973) administered a synthetic prostaglandin analogue intramuscularly and by infusion into the body of the uterus and similarly caused regression of corpora lutea more than 4 days old as well as induced estrus within 4 days. Oxender et al. (1975) found that mares treated by uterine infusion with $PGF_2\alpha$ returned to estrus an average of 2.2 days after treatment and stayed in estrus an average of 7.5 days (more than 2 days longer than control cycles). Thus prostaglandin administration terminates the luteal phase of the estrous cycle and returns the mare to ovulatory estrus.

Other Manipulations

Additional influences over the equine estrous cycle have been reported but not extensively investigated. Genital stimulation has been reported by Prahov (1959) to aid in inducing estrus in non-cyclic mares. In a study of the effects of rectal palpation on the reproductive characteristics of mares, Voss and Pickett (1975) noticed estrus lasted significantly ($P<0.05$) longer in non-palpated mares than in palpated mares. Furthermore, they found a higher percentage of non-palpated mares conceived earlier in the breeding season compared to palpated mares.

Nutrition and feeding patterns also influence the sexual behavior and ovarian activity of the mare. Belonje and Van Niekerk (1975) reported a study where seven mares were provided supplementary feed in the latter part of winter and another group had only natural pasture. In the supplemented group, all mares gained weight, showed estrus, and ovulated within 43 days; whereas among the eight non-supplemented mares, two lost weight and did not show estrus, four showed estrus but did not ovulate, and the two mares that gained weight had an ovulatory estrus. Mintscheff and Prachoff (1960) found that a feeding program of one day of no food followed by a controlled level of feeding on subsequent days caused a reduction in the duration of estrus.

[handwritten note in left margin: dominance plays role here?]

ABNORMAL SEXUAL BEHAVIOR OF MARES

From spring through autumn, mares typically exhibit estrus for a few days every three weeks; thus noticeable variations from this pattern are considered atypical or abnormal. Split estrus and prolonged estrus are examples commonly seen. In the peak of the breeding season, prolonged estrus is rare. During the spring and summer, some mares will fail to show clear signs of estrus yet may otherwise be cycling. Mares with physiological disruptions, including nutritional abnormalities, may not cycle normally; therefore, their sexual behavior will be altered. Some mares with congenital problems, such as abnormal number of sex chromosomes, have infantile reproductive organs and suppressed or irregular estrus.

Occasionally a mare may show stallion-like behavior (Figure 12.5). The in-

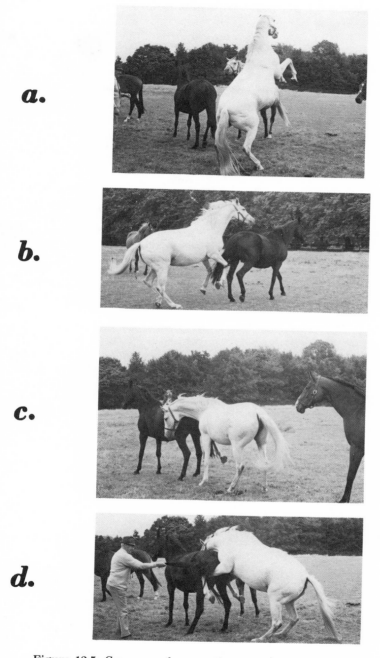

Figure 12.5: Grey mare three-months pregnant exhibiting unusual posses-siveness, teasing, and mounting behavior toward a mare in estrus. (Photos courtesy of P.D. Rossdale)

dividual may approach another mare with arched neck, vocalize, and further tease the other mare (Stabenfeldt and Hughes 1977). Mounting by estrous mares as well as by non-estrous mares toward mares in heat can also occur (Rossdale and Ricketts 1974, Fretz, 1977, Asa et al. 1979). Unlike cattle, such events are atypical in horses. Azzie (1975) reported stallion-like tendencies in mares with a subcutaneous implant of estradiol once the initial estrus subsided and the mares resumed anestrus.

Stallion-like behavior was associated with a masculinizing ovarian tumor (arrhenoblastoma), elevated serum testosterone, and low serum estradiol in the case reported by Fretz (1977). After the neoplastic right ovary was surgically removed, the mare's behavior returned to normal. Such tumors may not explain all cases; for example, the pregnant mare who displayed the mounting behavior shown in Figure 12.5 foaled normally and was subsequently normal (Rossdale and Ricketts 1974). In the case of the mare-mare mounting observed by Asa et al. (1979), the estrous mare who exhibited mounting was repeatedly ignored by the stallion who tended and mounted another mare in the enclosure. Asa and her co-workers concluded the mare's atypical response was perhaps a form of redirected behavior under the highly sexual situation.

13

Maternal Behavior

Parental care of foals is primarily given by the mother. Only occasionally is apparent protection of a foal exhibited by the dominant stallion or by a mare other than the mother. Stallions in feral herds have been seen on a few occasions to retrieve foals separated from the maternal band, and mares without foals sometimes show a tendency to shelter and protect young of other mares (e.g., Feist and McCullough 1975). In most cases, a mare keeps her newborn at her side and greatly limits the direct contact her foal has with other horses. Soon after parturition, the mother develops a close association with the neonate and thereafter provides it with its nutritional and protective needs.

PRE-PARTURIENT BEHAVIOR

The length of pregnancy in horses is approximately eleven months. In a study of 498 Thoroughbred mares in England, Rossdale (1967a) found the average gestation based on last service by the stallion was 340.7 days with a range of 327 to 357 days for 95% of the mares. In South Australia, Ropiha et al. (1969) determined the duration of pregnancy for 522 Thoroughbred mares based on ovulation to parturition ranged from 315 to 387 days; ten foals were carried for more than 12 months. The average gestation in their study was 342.3 days. Hendrikse (1972) found sightly shorter durations in smaller breeds of horses than in larger breeds; in total, the average gestation for horses in the Netherlands was found to be 340 days.

Environmental factors, such as season and nutrition, interacting with such factors as the sex of the foal and individual variation of the mare can affect the duration of pregnancy. For example, well-fed mares appear to foal slightly earlier than mares on a maintenance diet (Howell and Rollins 1951). Pregnancy is longer for foals conceived in early spring compared to conceptions resulting from late spring or early summer matings (e.g., Hintz et al. 1979). Nutritional effects alone are not responsible for this trend. Interesting also is the observation that fillies are born a day or two earlier than male foals (Ropiha et al. 1969). The age of the mare appears to have little effect on the duration of pregnancy. Although foaling can occur throughout the year, among free-ranging horses most births occur from mid to late spring (Feist 1971, Tyler 1972, Welsh 1975, Keiper 1975, Green and Green 1977, Salter 1978, Boyd 1980).

Throughout most of the period of pregnancy, the mare's behavior is not greatly altered until shortly before parturition. Since hormonal levels change and other physiological alterations occur during pregnancy, subtle behavioral changes may well be happening that have not been documented.

Growth of the two mammary glands begins about a month prior to parturition, and milk secretion may appear several days before foaling. As parturition nears, waxy material usually appears at the end of the milk canal on the now enlarged teats. Relaxation of the pelvic ligaments cause a surface depression on either side of the sacrum, and lengthening as well as swelling of the vulval lips occur one to two days before the foal is born (Rossdale 1967a).

Within one to four hours before parturition, the mare begins to show evidence of increased discomfort and restlessness. Sweating may be evident at the flanks and girth. If conditions permit, mares sometimes seek isolation by leaving their social group or by letting the group move away. Blakeslee (1974) reported the separation in free-ranging Appaloosa horses may be as much as 5 km and suggested that the more dominant mares may separate the farthest. Tyler (1972) noticed varying degrees of isolation occurred in the New Forest ponies. Some mares were well isolated, others foaled while yearlings or other horses were nearby, and still others gave birth close to busy roads with concomitant spectators. Collery (1978) concluded that young mares, especially those foaling for the first time, were the mares that showed the least tendency to withdraw from the social group. Boyd (1980) found no evidence that mares sought isolation under the feral conditons of the Red Desert of Wyoming.

Mares maintain considerable control over the time of foaling by being able to prolong the initial state of parturition if disturbed (Koch 1951). Thus my efforts and those of others to film and be on hand during parturition have been considerably frustrated because of the reaction of mares to bright lights and human observers. Mares tend to give birth during darkness or in the early morning hours when light levels are low and disturbances are greatly reduced (Rossdale and Mahaffey 1958, Flade 1958, Zwolinski 1966, Rossdale and Short 1967, Tyler 1972).

PARTURIENT BEHAVIOR

Having achieved some degree of isolation, a mare signals impending parturition by her restlessness, patchy sweating, and overall uneasiness. Circling, pawing, and occasional displacement eating may occur. Additional signs of discomfort appear with repeated recumbency and standing, looking at the flanks, tail raising, and restricted rolling (Wright 1943, Rossdale and Mahaffey 1958, Walser 1965, Rossdale 1967a). This initial stage of labor may last only minutes or occur for hours.

Rupture of the chorio-allanotic membrane and the escape of the allantoic fluid commences the second stage of labor. It usually occurs while the mare is standing preceded by vertical tail motions slapping the perineum and a crouching urination posture. Before becoming recumbent, the mare usually discharges noisily some of the allantoic fluid. Some mares investigate the allantoic fluid discharge and may then exhibit flehmen. A tendency to lick the fluid, their skin, and nearby objects is common, and sometimes a nicker is emitted as labor continues (Rossdale and Ricketts 1974).

Recumbency soon occurs and strong contractions become evident. A large quantity of allantoic fluid is often discharged as the mare's hindquarters first contact the ground. Sternal recumbency is maintained with early expulsion efforts. Bouts of strong uterine contractions force the forelegs of the foal into the vagina with its muzzle inserted between or adjacent to the legs. Oftentimes mares stand and change positions as the forelegs and muzzle protrude from the vulva covered by the amniotic membrane. Some mares get up and down repeatedly, especially if disturbed (Rossdale and Mahaffey 1958).

Final delivery of the foal is nearly always completed while the mare is in lateral recumbency with legs extended (Figure 13.1), rarely while standing. Numerous expulsion efforts occur. In a sample of five mares, Rossdale and Mahaffey (1958) counted 60 to 100 straining efforts occurred between the first appearance of the amnion until delivery was completed. Mares often elevate their extended upper hindlimb during expulsion efforts. Incomplete rolling movements sometimes occur. Much of the time is taken up with delivery of the head and forequarters then proceeds rapidly until the foal's hips are delivered and only the hindlegs remain in the vagina. At this point expulsion efforts cease (Rossdale and Mahaffey 1958) and the foal is considered born.

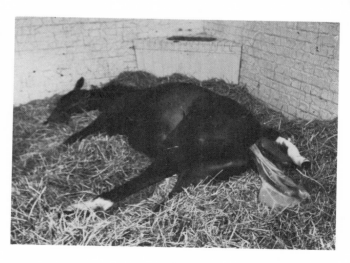

Figure 13.1: The appearance of the forelimbs and head of the foal during parturition. (Photo courtesy of P.D. Rossdale)

The interval from the rupture of the chorio-allantoic membrane to the delivery of the foal's hips varies from one occasion to another. Rossdale (1967a) found it averaged 18 minutes (range 5-47) for mares experienced in foaling and 21 minutes (range 5-43) for first-foaling mares. For 24 pony mares, Jeffcott (1972) found the second stage of labor averaged 12 minutes (range 4-25).

Parturition occurs with the foal usually encased in the amniotic membrane. Movements of the foal lead promptly to the rupture of the amnion and breathing commences. Normally this occurs when the foal raises its head away from the forelegs during delivery. The forequarters become exposed as the foal slides from the membrane (Figure 13.2).

Figure 13.2: When the mare's contractions cease, the foal's hindlegs remain in the mare and the umbilical cord remains intact. (Photo courtesy of P.D. Rossdale)

In the first moments after birth, the mare and foal are relatively inactive; the mare remains in lateral recumbency, and the foal assumes sternal recumbency. Unless distrubed, the mare typically remains recumbent for many minutes. As the foal begins movements to drag its hindlegs free of the vagina and membranes, the mare often assumes sternal recumbency and turns her head toward her foal. Nuzzling of the foal and quiet nicker vocalizations sometimes occur.

The umbilical cord does not sever until the foal moves itself away from the mother or until the mare attempts to stand (Figure 13.3). Rossdale and Mahaffey (1958) observed that the delay of several minutes before the umbilical cord breaks allowed time for the physiologically-beneficial transfer of 1000-1500 ml of placental-fetal blood to reach the foal. Severance of the cord too soon therefore results in depriving the foal of considerable blood otherwise left in the tissues of the afterbirth. Since mares normally remain recumbent, the foal's creeping movements away from the mother usually cause sufficient stretch to break the cord about 3 cm from the foal's abdomen. Thus several minutes after contractions cease, the foal commonly pulls its hindlegs free of the mother as well as the membranes and accomplishes severance of the umbilical cord.

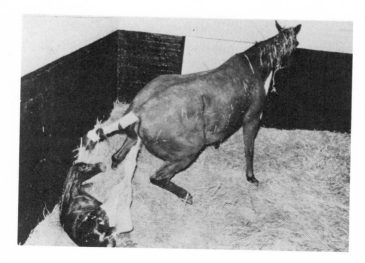

Figure 13.3: Foals often free themselves from the fetal membranes, but here the mare stands soon after parturition, pulling the membranes and breaking the umbilical cord. (Photo courtesy of P.D. Rossdale)

POST-PARTURIENT BEHAVIOR

While the foal struggles to achieve standing, the mare usually stands and soon begins a prolonged period of licking her newborn. Continuous licking may last 30 minutes. The vigorous licking proceeds over the foal's body and once complete rarely occurs again with such persistence. It is during this early contact that the mother commences her strong social attachment to her foal.

In human-attended barn situations, Rossdale (1967a) found 83% of the mares observed (n=257) stood within 16 minutes after foaling; some mares remained recumbent for up to 40 minutes. The presence of attendants may have induced some early standing; nevertheless, only 23% of the mares stood prior to 4 minutes after delivery.

Mares become protective of their foal during the early neonatal period. As intruders approach, sheltering and herding of the foal is exhibited. If the foal is lethargic, the mare often becomes impatient and may strike the foal gently with her forefoot. Stillborn foals are eventually pawed forcefully by mares (Rossdale and Mahaffey 1958). Once while we fondled a recumbent newborn foal in our experiments with human socialization, the restless mare gave the foal a swift strike with one foreleg as if to induce the foal to stand and withdraw with her.

When not attending the foal, a mare may nibble at hay or straw, smell and lick objects smeared with birth fluids, and become occupied with discharging the placenta. Expulsion of the afterbirth occurs on the average 60 minutes after delivery and concludes the third stage of labor (Rossdale and Ricketts 1974). By the end of the second hour, most mares (86% in Rossdale's 1967a study) have discharged the placenta; yet retention can last for hours. Tyler (1972) observed one free-ranging pony with a retained placenta 8 hours after delivery. Inconsequential retentions lasting up to 24 hours have been reported by Wright (1943).

Expulsion of the placenta normally is preceded by repeated sessions of recumbency with rolling and restless evidence of discomfort. The mare may sweat, look at her flanks, paw the ground, and groan. Meanwhile the foal has been able to achieve a standing posture. During the bouts of recumbency and rolling by the mare, the foal commonly exhibits excitement and circles or paws the mother. Maternal care is noticeably reduced during the peak period when the mare struggles with discomfort prior to placental expulsion. Final discharge of the placenta often occurs during recumbency or just as the mare stands.

After the placenta and membranes have been discharged the mare begins to calm. She usually smells and may even poke the material with her upper lip; a flehmen response often follows. A mare does not consume the placenta and only rarely has been seen to nibble the afterbirth. The horse is not adapted to hide foaling evidence. Under wild conditions, a mare and foal soon depart from the foaling site.

Once the foal stands, some mares position themselves to assist the foal in locating the teats. Nusing soon succeeds in such instances. However, other mares show avoidance instead and seem to resent their foal's activity near udders that are obviously quite sensitive. These mares may pivot and move away from the searching foal. In extreme cases, the foal may be bitten or kicked. If the foal places its head under the flank, the mare may squeal and bump the head of the foal with the stifle region of the hindleg during a hindleg lift. Avoidance may continue for hours; nevertheless, with time and continued care-seeking by the foal, initially obstinate mares begin to allow nursing.

A mare normally only allows her own foal to suckle. Smell, visual cues, and even auditory and gustatory cues seem to be utilized by a mare in recognizing her own offspring. The mare's recognition and attachment seem to initially develop during the licking and close contact of the first hour postpartum. Cox (1970) noticed that after months of separation only the mother and not other mares reacted with interest to a foal isolated from her several hours after parturition and subsequently hand reared.

Fostering a foal to another mare is usually difficult. One technique commonly utilized is to pair a foal recently orphaned to a mare that has lost her own foal (Tyler 1972, Rossdale and Ricketts 1974). The mare is induced to accept the strange foal by initially draping the foal with the hide of the mare's dead foal or with the amnion that covered her foal. The mare is allowed to smell as well as follow the disguised foal; nursing is encouraged by handlers. Successful acceptance of the foal by the foster mother will most likely occur in 1 to 12 hours. When the hide or amnion are not available to accomplish fostering, the strange foal's odor can be masked with an odoriferous ointment placed on the foal or on the mare's nostrils (Rossdale and Ricketts 1974).

Within a few hours, a mare leads her newborn away from the foaling site and together they rejoin social companions. Blakeslee (1974) concluded that subordinate mares rejoined their band sooner than dominant mares with their newborn foal. The mares were observed to move in bouts, stopping periodically. Sometimes herd members came to investigate.

A mare seldom willingly allows her newborn foal to have direct contact with other horses or humans. She calls the foal to her side with quiet nickers or intervenes with her body and herds the foal away. Previous young and strange foals are normally rejected using bite threats or kick feigning. When an intruder persists, the mare is apt to kick. Occasionally other horses attempt to adopt the newborn, especially mares that are soon to give birth. Blakeslee (1974) found geldings also showed the tendency to adopt foals. Boyd (1980) saw one instance where a deserted foal was adopted by a stallion and his band of five mares.

By withdrawal with her newborn sheltered at her side, a mare avoids most direct confrontations, such as with dominant individuals of the social group. If the foal is recumbent, the mare first rouses it with a quiet nicker or a nudge with her nose, causing it to stand and allowing timely withdrawal together. Although a subgroup in itself, the mare and newborn foal under free-ranging conditions are commonly a part of a larger social unit. The mare's close protectiveness and constant togetherness with her newborn is maintained for many days after parturition and is only gradually reduced.

A mare is most inclined to allow previous offspring, the more intimate mare companions (e.g., allogrooming partner), and trusted human handlers to first contact her foal. Protection and defensiveness may continue to be shown when other individuals approach. By four weeks of age, social interactions with other than the mother become more numerous, and foals increasingly interact in play and mutual grooming with peers (Tyler 1972).

The mother and foal both participate in maintaining periodic contact with each other. They separate only minimally. Tyler (1972) found, for example, that during the first week foals were within 5 meters of the mare 94% of the time. At five months, mother-young pairs still spent 25% of the time within 5 meters and less than 10% at distances greater than about 45 meters. Beginning with the eighth month, the common distance of separation was between 5 and 25 meters. Boyd (1980) observed similar tendencies in the feral horses she observed.

Nursing of the foal decreases from a rate of about four bouts per hour in the first week to approximately once every two hours when the foal is eight months old. Natural weaning occurs at about one year of age, shortly before the mare gives birth to a new foal (Tyler 1972). Some mares wean their foals earlier, and if the mare does not have a new offspring, nursing of the previous foal often continues. Only rarely do mares allow more than the most recent offspring to nurse.

Mares appear to initiate some bouts of nursing by approaching their foals and standing nearby. Boyd (1980) noticed mares with foals less than one week of age usually shifted weight on the hindlegs in a rocking motion before terminating a nursing bout. This pattern seemed to induce the foal to withdraw its head before the mare moved away. Mares can prevent nursing by walking away, bumping the foal's head with a forward lift of the hindleg, or by aggressive biting or kicking. At weaning, the mare may repeatedly drive the foal away as it approaches.

After weaning, the mare and her offspring maintain some degree of companionship that may last into adulthood or only until the offspring becomes sexually mature or departs from the original group (Tyler 1972). In some cases, youngsters depart at a year of age. Others do so later. And yet, some offspring remain in the maternal group as adults; oftentimes they can be seen to enter into a mutual grooming relationship with their mother.

Part V

Social Behavior

14

Social Organization

HERD STRUCTURE

Although some horses roam as solitary individuals, most horses prefer to remain with companions. Discrete social groups are called bands. A herd is a localized population consisting normally of one or more bands as well as solitary individuals. Interband dominance indicates that even at the level of the herd some social structure exists (Miller and Denniston 1979). Bands of over 20 animals occasionally occur. More often, however, the band size of free-roaming horses is less than 10, with 4 being the most common (Figure 14.1). In the arid Pryor Mountain Wild Horse Range of Wyoming-Montana, Feist (1971) found 44 family (harem) bands had an average size of 5.0 (range 2-21). Welsh (1975) observed up to 50 bands on Sable Island and found the average band size was 5.5 (range 2-20). By comparison, the feral horses Salter (1978) studied in western Alberta had an average band size of 7.7 (n = 18, range 3-16).

Several kinds of groups can be seen in a herd. Besides the typical family or harem bands consisting of at least one mare and her recent offspring plus an adult male, occasionally an additional male will accompany harem bands. Bachelor males often form small, less-stable assemblages of usually 4 or fewer members. Membership in bachelor groups commonly shifts throughout the year. Feist (1971) frequently saw solitary males as well as bachelor groups of up to 8 individuals; the average bachelor unit was 1.8 (n=23). Solitary mares sometimes are seen. Occasionally mares and offspring are together as a group without the company of an adult male. This is common when stallions are scarce. Another type of grouping that occurs is the non-family, mixed-sex peer band. Juveniles especially may assemble as a peer band and remain together for prolonged periods (e.g., Baskin 1976, Goldschmidt-Rothschild and Tschanz 1978).

Feral horse populations become structured into (a) reproductive components, consisting basically of harem bands, and (b) non-reproductive components, consisting of bachelor males, solitary females, and non-breeding juveniles or subadults (Figure 14.2). The population remains approximately half female and

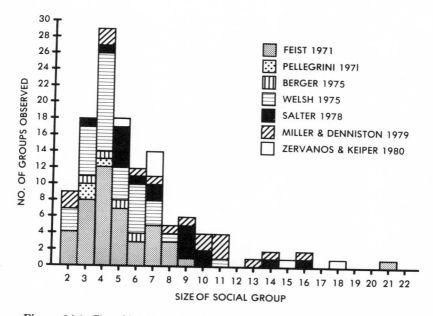

Figure 14.1: Size of feral horse bands observed in a variety of North American habitats. Data represent summer sample from one year of each study.

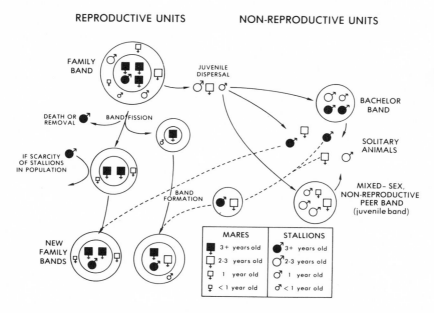

Figure 14.2: Social organization patterns exhibited by free-roaming horses.

half male, with adults outnumbering the yearling-foal age class by a factor of 3 (Figure 14.3). In Wyoming-Montana, a sample of 270 feral horses was found to consist of 48% males and 52% females (Feist and McCullough 1975); in Nevada, a similar-sized population of feral horses was composed of 47% males and 53% females (Green and Green 1977). The birth sex ratio is basically 1:1.

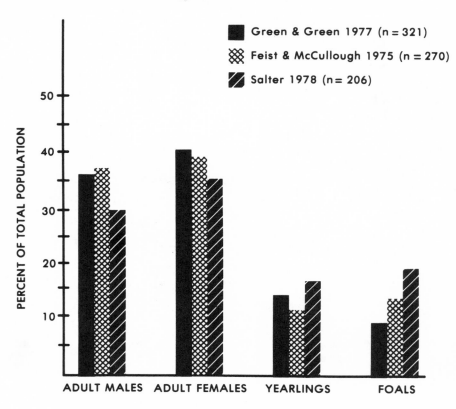

Figure 14.3: Age and sex composition of feral horse populations.

Mortality is greatest early and again late in life; nevertheless, horses have lived past the half century mark under management conditions. Under feral conditions on Sable Island, male life expectancy (5.85 years) was found to be greater than that of females (4.56 years), apparently because of reproductive stresses faced by mares (Welsh 1975). Feist and McCullough (1975) found mortality decreased after the first year of age and gradually increased again beginning about the tenth year. Small bands as well as unstable bands show the lowest foaling rates and survival of young (Welsh 1975).

The reproductive units are the mainstay of herd structure. Unless adult males are scarce, each harem is normally escorted by a stallion, occasionally by more than one. The typical harem consists of 1-3 mares and their offspring of the past 2 to 3 years. The nucleus of the harem band appears to be one or more adult females, not the stallion. Foals and yearlings show a preference to remain with their mother as well as with other young. And certain adult females show a mutual attachment and preference for each other's company. Thus these companions tend to remain together even without the presence of a stallion (cf. Tyler 1972). The stallion is, therefore, to some degree an adjunct social member of the mare subgroup. Stallions do occasionally show efforts to collect additional mares, and thus a harem band may contain more than one subgroup. Although the stallion is not necessarily the focal point of group affinity, he maintains a partriarchal position in the band and defends the band from intruders. Cooperation is shown the stallion by group members. Feist (1971) and Miller (1980) observed instances where harems seemed to delay their travel to facilitate the reunion of the harem stallion with the band when he was detained or otherwise separated from his group.

Because of herd social organization, most adult females but relatively few sexually mature males regularly participate in reproduction. The remaining individuals, including most bachelor stallions as well as juveniles and subadults are non-reproductive units of the herd. Further development and experience will eventually alter the status of the majority of these. Therefore, the life cycle and dynamic social organization of horses eventually provide reproductive potential to most individuals.

EMIGRATION AND IMMIGRATION

The tendency is for juvenile horses to disperse from their maternal band. Young males may remain solitary for months or even years or join other males in a bachelor group. Young females commonly join harem bands, but may for awhile remain solitary. Sometimes juvenile horses join a mixed-sex assemblage of other young horses that have also dispersed from their maternal band (Keiper 1976a, Goldschmidt-Rothschild and Tschanz 1978).

The age at dispersal is variable, apparently dependent upon prior experience and environmental circumstances, including social pressures (e.g., by adults of the same sex as the juvenile). In New Forest ponies, Tyler (1972) found many young dispersed by three years of age; by the time the ponies were four, most had dispersed. However, on Assateague Island, Maryland, Keiper (1980) noted that immature feral ponies usually dispersed before they were two years of age. The relatively low percentage of fillies that do not disperse retain membership with their maternal band. Although not common, some mares rejoin their maternal band after prolonged absence. Harem stallions under such situations react to these mares as they do to the other adult mares, including showing sexual interest upon estrus.

Except for juvenile dispersal and the occasional addition of a new mare, harem bands are relatively stable. Miller (1980) found the average change in membership in feral bands he observed was 0.75 adult changes/band/year. Emigration by adults is often only temporary. However, the death or removal of a key group member can cause band fission. The subgroups may then join other bands or merge with a solitary adult to re-organize a harem band.

The tendency in the non-reproductive social groupings is for the more mature individuals to eventually leave the assemblage. Typically a male consorts with a solitary female to establish a new band (Keiper 1980). For females, an alternative is to join an established harem. Some lone males may tag along with harem bands. Once tolerated by the stallion and mares, the subordinate male becomes a member of the family band and is a potential heir to the stallion's position should anything happen to the harem stallion. Some subordinates occasionally breed mares. Welsh (1975), Denniston (1980), and Miller (1980) noted that subordinate males help protect and maintain the band. Therefore, band size of multiple male harem bands is often larger. Miller found also more stability in such bands than even in harem bands having one stallion.

Established social units are not readily open to admitting strangers. Such is the case for harem bands (e.g., Feist 1971), bachelor bands (Salter 1978), and mixed-sex peer groups (Goldschmidt-Rothschild and Tschanz 1978). Age and sex of the approaching stranger can influence the response shown by group members. Harem stallions usually threaten and chase away approaching males. Sometimes serious fights occur between stallions. Occasionally stallions try to herd stray or solitary mares into their band. On other occasions, harem stallions have been observed to drive mares away from their band (Feist 1971). Foals are sometimes allowed to approach with little objection. Band mares sometimes participate with harem stallions in rejecting strange males. And in some cases, band mares also reject non-member mares who seek membership on their own or are invited by the harem stallion.

SOCIAL ROLES

A harem stallion often herds his band together upon the approach of other bands or intruders. If subordinate males are within the band, they commonly participate in herding and defense. One stallion noticed by Feist (1971) commonly encompassed an old stallion with a female companion into the herding of his band; yet, once the alarm was over, the old stallion and mare were allowed to stray several hundred meters away. Of the 130 instances of herding or driving by stallions observed by Feist, 42% occurred to move the group away from another band or stallion, 30% were to tighten or direct the movement of the group, 12% were attempts to copulate with a mare, another 12% occurred to herd away non-members from the group, and 4% were attempts to move non-members into the group.

Although stallions attempt to keep adult mares in their band from straying, Collery (1969) found the stallions he observed made little attempt to retain their own fillies. Stallions seem lax about the wanderings of their daughters at puberty; yet Feist (1971) noted one instance where a stallion retrieved an estrous filly. Harem stallions are typically not sexually motivated by the estrous displays of fillies, especially their own offspring, and generally limit their sexual interest to adult mares of their band. Estrous fillies are thus inclined to wander, apparently seeking male attentiveness. The excursion from the original social unit may be brief or in some cases permanent.

In populations where the number of stallions or their social activities are limited by man, the herding and defense role of the stallion in a band's social structure is assumed by a dominant mare who emphasizes the activities in the stallion's absence (Ebhardt 1957, Zeeb 1958, Tyler 1972). Thus under such situations herding and defense of social units may continue to be seen but are instead exhibited by the dominant female of the social group.

Although ultimately the most dominant member of a band can influence the group's activity, initiative and leadership are not shown only by that individual. An activity can be initiated by any member and it often then becomes a group activity through the phenomenon of social facilitation. Tyler (1972) observed that if a subordinate initiated a change in location it usually stopped and allowed a more dominant horse to pass and take the lead. In the feral horse bands observed by Feist (1971), the stallion was generally at the front of his band during travel. In 159 instances Feist observed, the stallion overtly took the leadership role in 122 (76.7%). When intruders were near, the stallion assumed a position between his band and the intruders. When a family band was alone and undisturbed, Welsh (1973) noticed harem stallions on Sable Island were basically passive and followed rather than led; the senior mare seemed to initiate most movement. Miller (1980) reported stallions and mares each led the band in about half of his observations; when trailing was observed, stallions were behind their band more often (73%) than adult mares (19%).

One of the major social roles of mares is attentiveness to the needs and welfare of offspring, especially the youngest. Maternal protectiveness is most intense with neonates and relaxes as the foal develops. During daily activities, mares and other members of the social group tend to maintain their location within the proximity of other members of their band. Foals may leave their mother temporarily to seek age-mates for play and mutual grooming.

15

Social Attachment

It is not unusual for horses to seek social contact with other horses; yet that social contact is usually directed toward specific herd members. Observations reveal social attachments or the bonds between individual horses are evident at various levels in band social structure. In effect, best friends pair off with best friends. Each mother and her young foal exhibit an intimate relationship; juveniles seek specific playmates; mares associate with only certain other mares; and even a stallion's associations are far from indiscriminate. Social companions arranged by man may not necessarily achieve much social unity even when pastured together if social attachments do not form (cf. Altmann 1951).

In horses, social attachments are the threads that hold social units, such as bands, together (Figure 15.1). Biological advantages of group living can then ensue. The stronger the interactive bonds, the more stable the social grouping and the more the individuals will function physically, temporally, and behaviorally as a unit.

The strength of bonds is affected by changes in reproductive condition, maturation, health, and other factors impinging on the individuals. As one relationship alters in a band, the change may ricochet throughout the social unit causing other alterations. In the extreme, such as when a key individual is no longer present, the disruption may even lead to group fission.

Once a horse has whatever social attachments are appropriate to its age, sex, and physiological condition, the drive to attain additional social attachments wanes. For example, once a stallion has a harem he rarely solicits new mares (Feist and McCullough 1975, Baskin 1976); a mare with foal is unreceptive to other foals (Tyler 1972). If a horse lacks an appropriate type of companionship, the individual often shows evidence of the social need and may actively seek or solicit a replacement to fill the void. For example, a stallion without a harem actively seeks mare companions; a foal without a foal playmate will usually solicit play and grooming from older horses; a mare or an orphan foal

Figure 15.1: Even when there is room to spread out, members of social groups remain close to each other primarily because of social bonds.

who has lost its respective partner is relatively receptive to a foster relationship; a mare without a foal may occasionally become highly protective of another mare's foal.

MARE-FOAL ATTACHMENT

A mare's attachment for her foal (mare-foal or maternal attachment) begins to be evident within minutes after parturition. She shows protectiveness and appears anxious if the foal is taken out of her reach. As the foal struggles in its efforts to stand, the mare, even while still recumbent, studies her newborn and utters a consoling quiet nicker as the foal collapses. As the foal nears her forequarters, the mare nuzzles it. With the approach of intruders, the mare stands and shelters the foal beneath her neck or at her shoulder (Waring 1970a,b).

Soon after first standing subsequent to parturition, the mare commonly grooms the foal with a prolonged bout of licking. At this time, the foal is still wet with amniotic fluid, and the sensory perception the mare gains during this grooming appears important in firmly establishing the mare's attachment to that particular foal. Subsequently, the mare discriminates between her own and other foals.

When the foal is not within reach after parturition, the mare's chemo-perception is directed instead to the fetal membranes and fluids at the site of parturition. During experiments when my co-workers and I removed foals temporarily from their mother following birth, the mares spent considerable time smelling and nuzzling rags that were used to dry the foal.

Upon a reunion of a mare and foal, the mare smells the foal apparently to test or compare sensory cues emanating from the foal to a memory trace of odors associated with perhaps the fluids of the recent parturition and the neonate itself. The mare is noticeably receptive to a foal having the odor experienced at parturition and shortly thereafter. After additional exposure to her foal, the mare begins to show evidence that she can also utilize visual and auditory cues in recognition, especially when the foal is not close by (e.g., Wolski et al. 1980). Smelling of the foal is still common once within reach.

Maternal attachment is especially linked to the post-parturient physiological state of the mare. Certain hormonal levels are probably involved. At parturition, a mare enters a sensitive period where the formation of a social attachment to a neonate is acute. Normally the mother's bond forms to her own foal; yet if the foal for some reason is separated from the mare at the early stage of the relationship, the mare may establish her maternal bond to another foal or to a surrogate. Once the bond is established, the sensitivity for establishing a new maternal bond wanes. If later the foal is lost or dies, the mare becomes depressed and socially disrupted. How overtly the mare shows her attachment to a foal appears directly proportional to the intimacy and length of time she has had with the newborn during the early postpartum period.

Fostering another foal to a mare is far easier soon after parturition than it is a day or two after the mare has affiliated with her own newborn. One of the more effective ways to get a mare to accept another foal subsequent to the establishment of her maternal bond is to drape the strange foal in the hide of her own dead foal. Tyler (1969) noted this technique was successful in an unusual case where a mare accepted an orphan foal after her own 3-month-old foal died in a road accident. Obviously, if either the mare or foal refuse to accept such a new arrangement, fostering fails. Both must be receptive.

Pain can disrupt maternal attachment. The formation of the bond may be slowed. When the bond is already established, the display of maternal behavior is affect by pain. Thus a mare in extreme discomfort may exhibit little evidence of maternal attachment or will be atypically forceful with her foal. Data are not sufficient to conclude whether maternal attachment differs significantly between multiparous mares versus first-foaling mares; differences in maternal attachment appear to be more the effect of individual mare traits rather than of foaling experience. Young mares as well as old can exhibit strong attachment to their newborn foals. The ability to form postpartum maternal attachments can last throughout a mare's life cycle. One mare I observed had her first foal at the age of 25; displays of her maternal attachment and maternal behavior were consistent with those seen in other mares.

A mare's attachment for her foal continues at considerable strength throughout the foal's first year. The first day or two the mare remains very close to the newborn and only gradually does she relax this protectiveness. As the weeks go by, the mare and foal separate for longer periods and at greater distances (Figure 15.2). Nevertheless, Tyler (1969) found even at 5 months of age, foals spent less than 10% of the time more than about 50 meters from their mother.

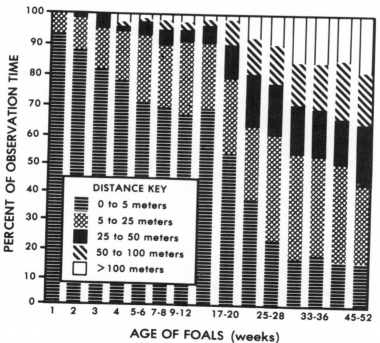

Figure 15.2: Change in the distance between foals and their mother during the first year. (Data from Tyler 1969)

Toward the end of the first year, the maternal bond is generally still evident, but the intimacy between the pair is noticeably reduced. Weaning usually occurs in the few weeks or days before the mare is to foal again. With the subsequent parturition and arrival of a new foal, the mare's attention and social activity shift abruptly to the neonate. She tries to remain relatively isolated with her newborn. The close approach of the previous offspring is discouraged by the protective mare for the first day or two. Within a few days after parturition, the mare and previous offspring may be seen to graze and rest near each other and engage in mutual grooming. Seldom is the previous young allowed to resume nursing. Although the mother's bond to the previous young seems to change most drastically at the arrival of the new sibling, the previous maternal bond may last for years, especially with female offspring, as evidenced by later mother-offspring interactions and affability. The relationship may continue even after offspring begin to have foals of their own. Tyler (1969) found one 13-year-old mare still groomed with her mother even though they were in separate social groups.

If a new foal is not born the next foaling season or for some reason the mare does not have a newborn at her side, her attachment to the prior offspring remains very evident and in some cases seems to strengthen. In the latter cases, mares may maternally interact in a more intimate way with their yearling or older offspring than would otherwise occur. For example, nursing may continue and protectiveness may be evident. Shelter seeking by older offspring toward the mare, maintaining proximity, and mutual grooming between mare and older offspring may occur whether the mare is accompanied by a newborn or not.

Sometimes mares without foals of their own adopt newborn foals. Dominant mares and even geldings have been observed to steal a newborn from its submissive mother (Blakeslee 1974). The foal rarely survives unless the adoptive parent lactates and cares for all the needs of the foal. One newborn foal observed by Boyd (1980) was deserted by the mother soon after parturition when it would not follow with the maternal band; the foal was then adopted by an alien band of adults, consisting of a stallion and five mares. The foal was unable to obtain milk yet attempted to feed on plants by the age of two days. The youngster stayed with the foster band for several days; however by the twelfth day after parturition, the foal was no longer seen and was presumed dead. Eldridge and Suzuki (1976) concluded that in a case they investigated a mare mule adopted the first of twins born to a Shetland mare. The mule apparently came into lactation spontaneously and successfully raised the adopted foal.

FOAL-MARE ATTACHMENT

A foal's attachment for its mother (foal-mare attachment) normally begins after the maternal bond has already become established (Figure 15.3). And although difficult to quantify, the intensity of the mare's attachment for the foal seems greater than the foal's reciprocal attachment for her, at least in the first weeks after parturition. In general, the behavior of the foal appears more opportunistic.

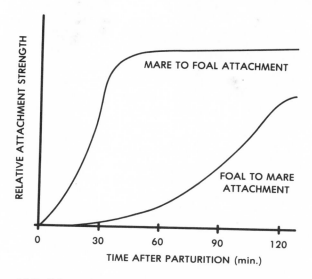

Figure 15.3: Schematic representation of bond development between mare and newborn foal following parturition.

Although the foal's eyes, ears, and chemoreceptors seem to function at or soon after birth (cf. Rossdale 1967a), the foal initially spends little time investigating its environment. However, around 25 minutes after parturition the foal begins to show distinct binocular orientation with the accompanying head movements. Subsequently, within another 10 to 20 minutes, auditory investigation with independent ear orientation becomes apparent. And finally, nosing, smelling, and licking of nearby objects becomes overt towards the end of the first hour after birth (Waring 1970a). These sensory experiences normally expose the neonate to its mother, who also provides input by licking, nuzzling, and emitting quiet nickers and weak whinnies. The foal exhibits care-soliciting behavior during its increasingly vigorous pre-nursing investigations.

Social interaction with the mother becomes especially obvious beginning in the second hour postpartum. By then the foal has stood and nursing usually occurs. Evidence of the foal's growing attachment for the mare is seen in the nuzzling and attentiveness the foal gives the mare as well as its efforts to follow and maintain a position close to her. As intruders approach, the foal begins to seek shelter close to the mare. When the mare becomes recumbent before discharging the placenta and groans in discomfort, the foal may circle her attentively and nuzzle her head and forequarters. In addition, the foal may respond to the mare's vocalization with a weak whinny or other sound (Waring 1970b).

As the relationship continues, the foal's attachment to the mare appears to strengthen. The longer and more intimately the foal associates with the mare, the more it shows a desire to be exclusively with her, and the less likely it will seek a relationship with other organisms. The senstivity of the foal to this initial social attachment occurs especially in the first few hours after birth. Fear responses to strangers become increasingly evident after the second hour postpartum if the foal-mare attachment has begun to develop. The foal's initial tendency to follow large moving objects becomes more and more restricted, becoming a response shown only to the object of its social attachment. Visual cues especially aid the foal in locating its mother (e.g., Boyd 1980).

Tests for distress behavior as a measure of foal-mare attachment have not been completely revealing. Although mares show overt excitement and anxiety, many foals when temporarily separated from their mother in the first few weeks show restless behavior more akin to disorientation than distress. Houpt and Hintz (1981) noted the vocalization rate of experimental foals temporarily separated from their mother peaks toward the end of the first month, suggesting foal-mare attachment is greatest at that time.

Both the maternal bond and the foal-to-mare bond are important for the well-being of the foal and help keep the pair together. The foal-to-mare bond limits spatial separation and establishes the foundation for later social development. It provides the foal with a source of nutrients, shelter, security, and guidance. The reciprocal mare-to-foal bond assures adequate parental investment to maintain close protection of the foal, to prevent spatial separation, and to provide for the needs of the foal. It is by a combination of the two bonds that the foal survives and may itself someday produce offspring.

Malfunction of the social attachment between foal and mare can occur. Tyler (1969) reported some foals born in the woodlands of England inappropriately form their attachment to a tree. In one instance, the foal showed considerable attraction to a particular tree on the day of birth. It remained close to the tree even when the mare moved several meters away. The foal nibbled the tree, walked around it, and moved to and from it. In cases where the foal completely ignores the mare, the mother eventually deserts it. Boyd (1980) reported such an example. However, in the case Tyler observed, a successful foal-mare relationship eventually formed.

Alteration of the social experiences of newborn foals can affect normal social attachment and subsequent behavior. Grzimek (1949a) once reared a foal in isolation from other horses until 64 days of age. His experiment gave early insight that social imprinting does exist in horses and that early social attachment and experiences greatly affect later social behavior. The foal at 64 days avoided and showed fear of other horses; it made every effort to remain close to human handlers. The foal's attachment was to people not to its own species.

Foals isolated from other horses and reared on a system whereby milk is mechanically dispensed also show defects in normal social behavior. They prefer human caretakers to equine companionship when later tested and fail to interact with the social signals of their own species (Williams 1974).

How maternal and social deprivation during development affect adult behavior, such as reproductive behavior, is not yet clear for horses. The effect may be major if the individual identifies with a foster species as has been shown in other farm animals by Sambraus and Sambraus (1975). Complete disruption would be unlikely if the foal had established some early attachment to its own species. Blakeslee (1974) reported such a case. A foal orphaned at eight days of age was isolated from other horses and hand-reared. Although it developed an attachment to its new human mother, the colt successfully joined into a group of yearling males when released later.

PEER ATTACHMENT

The social attachment that occurs between young horses causing play groups or that often develops between mares to form best-friend relationships is peer attachment. These relationships normally develop between horses of similar social or age class, hence the term "peer."

Peer attachment normally does not develop in foals until a number of days after birth. The protectiveness of the mare and the reluctance of the foal to approach other horses initially limit social contact to the mother-young relationship. During the first two to three weeks, foals seldom interact with other foals, and when they do, they mostly stare at each other or may briefly touch. Only after their third week, Tyler (1969) noted, do foals begin to move further from their mother. Interactions with other young then become more frequent. Early interactions are primarily investigative with approaching, sniffing, and nibbling. Progressively foal-to-foal interactions become playful. The foals chase one another, often rearing, kicking, and bucking as they gallop. In quiet moments, the foals begin to spend considerable time grooming each other.

As mutual grooming and playful interactions develop between foals, it becomes increasingly evident that the foals are developing social attachments outside of the mother-foal relationship. Foals progressively seek peer companionship. And although several potential partners may be available, foals pair off more and more with a particular partner. The companions often remain close during grazing, resting, and other activities.

The peer relationship that develops can be with a sibling or with another foal either within or outside the social band. Occasionally trios form. Partnerships can be between colts, between colt and filly, or between fillies. Colt partners alternate mutual grooming with long bouts of play fighting. Filly-colt and filly-filly partnerships are characterized by mutual grooming almost exclusively, with only rare interludes of playful chasing (Tyler 1969).

Bonds that develop between immature horses may or may not persist into adulthood. When juveniles disperse from their original companions, previous social ties often appear to be severed; yet, long-term information is sketchy. Perhaps previous ties promote social contact and renewed companionship later, especially among mares.

Some immature horses disperse from their family group in the company of a companion rather than alone. The pair may then join another social group. Observations where lone females have been affiliated with a group of males (e.g., Feist 1971, Salter 1978) appear to be the result of a young female joining an existing male group in the company of her previous colt companion. Likewise, mixed-sex juvenile bands (e.g., Keiper 1976a, Goldschmidt-Rothschild and Tschanz 1978) may form, in part, by juvenile pairs joining an established juvenile group or serving as a nucleus for the formation of a new group.

New social contacts provide an opportunity for new peer attachments to develop. The ties apparently remain weak among males in a bachelor group, since such groups are normally not very stable. However, other social groups by comparison are more long term, suggesting better developed peer attachments. Regardless of group affiliation, the tendency is for non-reproductive individuals to eventually shift their social situation until they become a reproductive participant in the herd social organization. To achieve such a status may require more than one change in social group and corresponding changes in social bonds. Mature stallions tend to eventually shun male companions. For stallions, the overt social attachments once with a harem are primarily heterosexual and paternal relationships. Mares, however, can develop and maintain close relationships with other mares without interfering with reproduction.

Throughout the life cycle, mares tend to pair off with other mares. In adulthood, such peer attachment may be a carry over from previous juvenile companionship or more often is a relationship that develops anew in maturity. Some mare-mare companionships can be mother-offspring pairs that have persisted when female offspring remain with the maternal band or have rejoined it. In most cases, however, mares tend to pair with mares of similar rank and age (Wells and Goldschmidt-Rothschild 1979). Harem bands commonly are stable because of well-developed peer attachments. In a band, a mare may exhibit attachment to all or several peers; yet, it is not uncommon for mares in a group to exhibit a single best-friend relationship.

Mare companions tend to remain together in most activities. They graze together, rest together, and groom together. When separated, they become distressed and whinny loudly in an effort to make contact and become reunited. If kept apart, the desire to rejoin a companion may persist for many months. Tyler (1969) observed that a mare separated from her companion in the fall became reunited with the companion when released in the spring even though the remaining mare had joined with another group during the winter.

HETEROSEXUAL ATTACHMENT

Attachments between horses of opposite sex primarily for interactions of a sexual nature are heterosexual attachments. Often the partners interact as if in a peer relationship; yet underlying the relationship is sexual attraction. Few studies have focused on these bonds in horses. Although sexual promiscuity often characterizes horse management situations, the social system that typically develops under free-roaming conditions concentrates sexual behavior specifically toward companions within the band. Both stallions and mares appear to choose their sexual partners; even under management conditions, considerable biases and preferences for mates are shown. Mutual attraction facilitates the development of long-term heterosexual attachment. The consort arrangement between mare and harem stallion becomes more overt when the mare is in estrus because of the stallion's increased attentiveness and closer proximity. Nevertheless, the bond persists between successive estrous periods and likely began when the partners first accepted each other as band members.

Heterosexual attachment may begin as peer relationships among first year foals. Mutual grooming between a colt and filly is common, Tyler (1969) observed; but often the colt becomes rough and the filly withdraws. The colt sometimes follows and initiates further grooming, but soon roughness may again cause the filly to avoid the colt. In some cases, a filly returns to a colt, initiates grooming, nibbles at his head and legs, and in a playful manner rears or initiates chasing. Such colt-filly pairs often establish an intimate relationship and remain very close to each other. When another filly approaches the pair, the colt at most only sniffs her, whereas the filly companion usually threatens the intruder. When another colt approaches, the two males occasionally interact in a brief challenge before the colt returns to his companion.

Relationships involving first-year foals usually lack sexual overtones seen in older horses. Mounting attempts by first-year colts are feeble and rarely associated with erection of the penis. Mounting is not unique to colt-filly relationships; colts mount any companion, including the mother. When penis erection does occur in colts, the individual is usually resting, play fighting, or engaged in mutual grooming. Fillies exhibit no overt sexual displays until their first estrus as yearlings.

At puberty, social contacts of young horses often shift, and sexual behavior begins to characterize interactions between horses of the opposite sex. Dispersal from the maternal band often occurs at this time. New male-female encounters are often brief, with the sexual advances of the male being most obvious. Unless in estrus, the female reacts negatively to direct contact with a stallion, except for mutual grooming. Young females when in estrus solicit to males, including mature stallions and sometimes to first-year colts. Harem stallions seldom show an interest in the young mares, especially from their own band, and seldom tolerate sexual activity by nearby males, thus young females and males usually disperse. It may take months or even years for the young individuals to find a compatable companion of the opposite sex.

Once a mare and stallion mutually accept the presence of the other, their relationship can develop into a long-term heterosexual bond. Mutual grooming and sexual interactions seem to facilitate bond development. Body conformation, coloration, and even mannerisms may be important factors in sexual attraction and therefore may be involved in bond formation. Once the bond forms, the pair remain together spatially and socially throughout the year. In some cases, a solitary stallion locates a receptive mare, and the two start a new band. Sometimes a mare joins an established harem band after gaining the interest of the stallion and the tolerance of the other group members. And occasionally a stallion will join a mare band that either has no harem stallion or where he has been able to supplant the previous harem stallion.

In most cases, heterosexual bonds develop between the stallion and all adult mares in the band. Cases where a stallion drives off a mare from his band (cf. Feist 1971) are possibly instances where heterosexual attachment did not develop. More often, mares are coveted and defended by the stallion. The stallion retrieves mares when they wander too far away, places himself between the harem and intruders, and often drives the harem to safety. Mares exhibit their mutual attachment for the stallion by facilitating and accepting heterosexual social contact with him only (not with bachelors or nearby harem stallions), awaiting his return when he has been detained, and responding at a distance to his vocal inquiries (e.g., Feist 1971, Baskin 1976).

PATERNAL ATTACHMENT

Harem stallions associated with family units normally extend their herding activities and protectiveness to young as well as to the mares within the band. Stallions sometimes retrieve foals that have wandered and can be seen to protect foals from pending danger (e.g., Feist 1971, Boyd 1980). Foals more than three weeks of age seem interested in stallions; they nibble them, and young colts even play fight with the gentle and indulging stallions (Tyler 1969). Thus there is evidence that the stallion has some paternal attachment toward offspring. Mares most often handle care and protection of their own foal and seldom does the need arise for the stallion to intervene; yet when the stallion herds or drives his harem, the young of the band are included in the stallion's maneuvering of the social unit.

As the young in a band mature, the harem stallion reduces his overview of foal activities. By the time a colt or filly reaches sexual maturity, the stallion passively allows dispersal. And in some instances, especially with colts, the stallion aggressively encourages a juvenile to leave the band (Jaworowska 1976). Few juveniles return; yet occasionally dispersal is temporary or does not occur, and the stallion may subsequently allow a young mare or submissive colt to return or remain as a member of the family band.

INTERSPECIES ATTACHMENT

Interspecies companionship occasionally exists in horses under management conditions (e.g., Olberg 1959). In such cases, a horse usually shows an attachment for a particular companion animal whether it is a chicken, goat, dog, human being, or other organism. Interspecies companions commonly replace peer relationships normally shown to other horses. The foal Grzimek (1949a) raised in isolation from other horses exhibited a generalized preference for familiar human handlers. A dependency on having the companion nearby may develop. In the absence of the companion, the horse involved tends to exhibit restlessness or signs of social deprivation. A race horse is considerably more relaxed when with its companion than without it, and owners are cautious not to loose the possible effect the companion has on the horse's well-being.

Interspecies companionships seem to develop in horses as a result of experience with the exotic animal during a prolonged situation where horse companionship is not available. The interspecies companion, therefore, becomes a surrogate for equine companionship and subsequently a long-term bond develops.

16

Home Range and Territoriality

Horses restrict their movements to a specific home range as well as to a limited distance away from social companions. Furthermore, horses show a tendency to defend an area around themselves and their social unit. Age, sex, physiological state, social status, and environmental situation appear to influence home range, social distance, and the somewhat subtle forms of territoriality. Considerable plasticity seems to exist in the species; thus population characteristics, home range size, and territoriality can vary noticeably between two study sites (Table 16.1).

HOME RANGE

The home range of an individual horse normally coincides with the home range of other members of its social group. It is the geographical area covered during day to day activities. The major requirements within the home range are water, food, and shelter. Shelter includes shade, wind breaks, and retreats from such things as insect pests. These resources may be shared with horses of other social units, thus the home ranges of more than one band may overlap.

As seasonal changes occur in the abundance of food, water, insects, and the need for shelter, the movements and habitat utilization patterns of horses adjust as well. Thus in times of abundance of food and water, little movement may be required to satisfy the needs of the horses. Or as happens in southern Nevada after heavy rain or snowfall, bands shift temporarily to areas of abundant food until surface water dries up, then the horses return to the portion of their range which contains a permanent water source (Green and Green 1977). Elsewhere, when food becomes scarce near water holes, horses may have to forage in one area and travel as much as 16 km to obtain water (Feist and McCullough 1976). Ponies that seek shallow bays or ocean surf during the insect season subsequently avoid entering the water during winter (Keiper 1979a).

159

Table 16.1: Population Characteristics and Social Patterns of Feral Horse Populations

	Wassuk Ridge (Pellegrini 1971)	Pryor Mountains (Feist 1971)	Grand Canyon (Berger 1975)	Sable Island (Welsh 1975)	Shackleford Banks (Rubenstein 1978)	Stone Cabin Valley (Green & Green 1977)	Boreal Forest (Salter 1978)
Population Characteristics							
Population size	12	270	78	240	104	703	206
Population density/km²	0.1	2	0.2	6.3	11	0.5	1
Age Structure:							
Adults, %	75	58	66	64	61	77	53
Juveniles, %	17	29	23	21	21	14*	29
Foals, %	8	13	11	15	19	9	18
Activity Areas:							
Group home ranges (average), km²	31	15	23	3	6	52	8
Group territories (average)	?	no	no	no	3 km²	no	no
Social Patterns							
Group Composition:							
Harems							
Average size	3.3	5.0	4.5	5.5	12.3	5.3	7.7
Size range	3-4+	2-21	3-6	2-8+	–	2-15	3-17
Bachelors							
Average size	?	1.8	1.8	–	2.6	?	?
Size range	1-7	1-8	1-3	1-?	1-?	1-4	1-6
Mixed-sex peer group	?	yes	yes	?	yes	?	yes
Solitary males	yes	yes	yes	yes	yes	yes	yes
Band Stability:							
Adult female group changes, %	0	7.6	0	some	10.8	1.5	6.8

Modified from Rubenstein 1978

*Yearling age class only.

Home ranges of free-roaming horses vary greatly in size and are correlated not with group size but with resource availability. Home ranges vary, for example, from 0.8-10.2 km² (200-2500 acres) in the New Forest of England, from 2.6-14.4 km² in the boreal forest of western Canada, from 0.9-6.6 km² on an island habitat off Nova Scotia, and from 3-78 km² in arid regions of western United States (see Table 16.2).

Table 16.2: Variation in Home Range Size, for Harem Bands, Bachelor Bands, and Solitary Males at Several Locations

Home Range Size (km²)	Location	Source
. . . . .Harem Bands		
0.8–10.2	New Forest England	Tyler 1969
2.6–14.4	Alberta Canada	Salter 1978
0.9–6.6	Sable Island Canada	Welsh 1975
2.2–11.4	Assateague Island, USA	Zervanos & Keiper 1980
3–32	Wyoming- Montana, USA	Feist 1971
8–48	Arizona USA	Berger 1977
11–78	Nevada USA	Green & Green 1977
17–33	Nevada USA	Pellegrini 1971
. . . . Bachelor Bands.		
12.4	Alberta Canada	Salter 1978
. . . . .Solitary Stallions		
4.7	Alberta Canada	Salter 1978
5.2	Nevada USA	Pellegrini 1971

In most home ranges, resident horses rarely use all parts of the range equally. Some areas are used extensively and other sites are seldom visited. Thus bands have smaller areas within their home range where they spend most of their "Core area" time.

Rhythmicity in activity patterns and habitat use sometimes is very apparent. In such cases, horses tend to carry out maintenance behaviors at nearly the same time and place within the home range from day to day (Tyler 1969, Welsh 1973, Rubenstein 1978). Some bands may migrate seasonally within a larger range and utilize areas cyclically year after year. Yet in most bands, unpredictable and irregular patterns do occur periodically, and some groups seem to have little or no schedule for their site visits and activities.

Occasionally an individual or even a group will leave the previous home range boundaries and take up residence in a new range. The reason is not always apparent. Young horses that commonly disperse from the parental band as yearlings or older face a shift in home range; sometimes, however, they join a band whose range partly overlaps that of the original band. Some horses shift home ranges repeatedly. Old and decrepit horses may account for some of the shifts (Feist and McCullough 1975). Sometimes mares shift at the time of foaling. In other cases, an entire band may shift its home range because of human-caused environmental disturbances, such as seismic testing (e.g., Welsh 1975). Thus the geographical area any individual horse uses during its lifetime (life range) is considerably more than its home range at any particular stage in its life.

If a horse is taken from its home area (whether a stable or open range), it exhibits a tendency to return home once released (cf. Grzimek 1943b, Williams 1957, Tyler 1969). Some horses have successfully returned over distances of 15 or more kilometers. Homing succeeds in returning an individual not only to familiar habitat but also to its social companions. Undoubtedly, stress is then considerably reduced. Tyler (1969) observed an instance where a yearling colt was taken about six kilometers from his mother and previous range. After castration the colt was released. Five days later he was back with his mother on the original home range. Ponies allowed to graze on the Chincoteague National Wildlife Refuge in Virginia are annually rounded up and herded from Assateague Island to Chincoteague Island where culls are made for public auction. Subsequently, the remaining ponies are herded back into the water and released. Keiper (1979b) found that the ponies not only re-form the same social groups but they also re-occupy the previous home ranges.

During day to day activities, horses tend to move only a short distance from companions before moving back toward them or awaiting their arrival. This spatial limit is called social distance. Several factors, such as pending danger or strength of social attachment, affect the limit seen from one occasion to the next. In the feral population Feist (1971) observed, the social distance between members of a band was seldom more than 23 meters. Baskin (1976) found separations of as much as 50 meters but only under non-alarmed conditions. Social distances are the least between mares and their foal of less than a week of age. In 94% of the observations Tyler (1969) made of foals in their first week, the spatial separation between mother and foal was less than five meters. As foals mature, the social distance increases.

Early experience of a positive and diverse nature seems to boost the security and confidence level of young horses, thus the distance they separate from mother or peer companions can be relatively large. The foals we had given extensive early handling were very inclined to drift away from the mother in their eagerness to explore. The tendency for foals to normally limit their activities to a circle around the mother was emphasized by Baskin (1976). As the band moves to a new site, the mares move ahead of the young; but once grazing resumes, the youngsters drift somewhat ahead of the group before circling the mares in their activities.

TERRIT⬤RIALITY

In most feral horse herds that have been studied, territoriality or defense of an area against conspecifics is typically to prevent intrusion of a zone around one or more individuals rather than defense of a fixed geographical site. Most horse habitats do not have abundant and even distribution of all resources, thus it is common for horses of one band to share watering, feeding, and shelter sites with other bands. Overlap of home ranges of different bands is frequently observed (Feist 1971, Tyler 1969, Welsh 1975, Green and Green 1977, Salter 1978, Miller and Denniston 1979). Yet, to the extent possible, bands mutually avoid each other. Terrain can assist in isolating bands, such as the parallel ridges and deep valleys that Pellegrini (1971) found in western Nevada.

If one band encounters another, any defensiveness shown usually appears to be an attempt to maintain the integrity of the band rather than to defend a site. To obtain access to a resource, however, one band may attempt to displace another band and encounter some defensiveness from the band already at the communal resource.

Dominant stallions are often the individuals active in aggressive encounters between bands; however, if the stallion does not take the initiative other members of the band may, such as a dominant mare or subordinate stallion (e.g., Miller and Denniston 1979). Feist (1971) reported that the feral bands he observed usually maintained a spatial separation of 100 meters or more, and around water holes an approaching band would wait at a distance until an earlier group had departed. Subordinate bands tend to avoid more dominant bands; a dominant band can, therefore, displace a subordinate group or individual (Berger 1977, Miller and Denniston 1979). Marking dung and urine appears to enable a stallion to advertise his presence and may facilitate spatial distribution; overt marking of range boundaries per se has not been observed.

Not only do horses defend a zone around their band, but also each individual horse maintains some degree of personal space (individual distance) around itself, and a mare typically challenges those who approach her foal. A mare's defensiveness associated with her foal is greatest in the neonatal period and wanes as the foal matures. As horses develop, they begin to show a tendency to keep at least some distance between themselves and their nearest neighbor. Feist and McCullough (1976) noticed that threat displays were shown if one horse came within 1.5 m of a horse that would not tolerate close association at that moment (Figure 16.1). Individual distances appear to vary with the sex, age, social status, environmental context, and mood of the interactants. Personal space in horses seems to center on the head or forequarters; yet research is needed to determine the complete contour of personal space around a horse's body and factors that affect it. Individuals sharing a strong social attachment may exhibit little, if any, individual distance to each other under most circumstances. Oftentimes spatial distribution within and between bands is maintained not by overt defense but instead by avoidance. Baskin (1976) noticed the spatial separation that occurred between individuals within a band was typically 5-6 m; whereas between groups that shared a common feeding site, the separation was 40-60 m.

Figure 16.1: Head-extended threat display often shown by a horse in an effort to maintain its personal space. (Photo courtesy of P. Malkas)

Territoriality, social attachment, and social dominance are often interrelated. For example, a dominant animal socially attached to a group will defend the space around that group; yet if social attachment is weak or dominance status low, evidence of territoriality may also be poor. Harem stallions are faced not only with defending their mares and maintaining the integrity of their band but also with retaining their dominant social status. Bachelor males seek opportunities to breed with mares, to herd them away, or to supplant the harem stallion.

Another example of the interrelationship that can occur between territoriality, attachment, and dominance is seen when occasionally a mare attacks a stallion when he mounts her peer companion. Peer attachments often occur between horses of similar rank. Peer companions can display a degree of defensiveness, if not possessiveness, with regard to their social partner.

Certain environmental situations seem to facilitate the isolation of one band from another onto exclusive-use home ranges (Gates 1979, Zervanos and Keiper 1980) or onto territories *per se* where defense of the site is shown (Rubenstein 1978). In most habitats where feral horses have been studied, exclusive-use home ranges do not occur; however, it is not unusual for bands to utilize different core areas (areas of intensive use). In arid environments, two or more bands may utilize a common water hole as well as some of the same feeding sites; the bands remain spatially separated, yet their home ranges overlap. When resources are abundant and evenly distributed in a habitat, home ranges need not overlap. Bands may then partition the environment into adjacent exclusive-use ranges. Subsequently, each band can limit its movements to its own area and may even defend that area against encroachment by other bands.

The narrow barrier islands along the eastern coast of the United States seem to provide habitat where bands can establish and maintain exclusive-use areas. On Assateague Island, Maryland, for example, Zervanos and Keiper (1980) noticed some home range overlap when all data were plotted; however, when the infrequent excursions were excluded from the distribution plots, they found that bands maintained themselves in separate and adjacent home ranges arranged sequentially along the length of the island. It appeared as if each band might be occupying a territory, but no overt defense of the sites has been seen. Nevertheless, Rubenstein (1978, 1981) has seen overt defense of exclusive-use areas on Shackleford Banks off the coast of North Carolina; the defended areas provide all the needs of the resident band. The population is more dense than most horse habitats (Table 16.1). Two-thirds of the harem groups maintain the well-defended, non-overlapping, permanent territories spaced sequentially along one end of the elongated main island.

Territorial boundaries on Shackleford Banks seem to coincide with subtle geographic features, such as a patch of fresh water, a tidal inlet, or a row of low sand dunes. Large dung piles are distributed throughout the territory and do not appear to be boundary markers. Boundaries run the width of the island from the ocean to the sheltered back waters. The boundaries of the approximately 3 km² territories have shifted no more than 15-20 meters over several years of observation (Rubenstein 1978 and personal communication).

As soon as a territorial stallion on Shackleford Banks detects that an intruding male has entered his domain, he charges and a fight normally ensues. The resident invariably wins and the intruder retreats. Most of the time, territorial stallions and their harem have little outside interference or competition; the overall energy cost of maintaining a territory on the island seems quite low. The narrow island reduces exposed boundaries and visual monitoring of the territory is easily accomplished. Only from May through July do males sometimes raid territories apparently in search of mares; otherwise bachelor males and neighboring stallions with harems rarely cross boundaries of territories. A notable exception occurred during an extreme drought. Twice a stallion herded his harem across the territorial line to drink from a neighboring water hole after theirs had become dry. During each excursion, the stallion restricted his band to the sand flats and proceeded only at low tide (Rubenstein, personal communication).

When the behavior patterns of territorial bands are compared to non-territorial harem bands, additional differences appear. Territoriality seems to provide some adaptive advantage for maintenance activities and for reproduction. Territorial bands on Shackleford seem to exert less grazing pressure on the various patches of vegetation within their exclusive-use range, returning to a patch every 10-14 days compared to less than 7 days for bands with overlapping ranges. Territorial harems consist not only of more adult females but also have been the only groups that have consistently shown a size increase. Non-territorial bands, possibly in response to more frequent contacts with other bands, show smaller average individual distances during and immediately prior to the breeding season. Territorial stallions spend relatively little time driving their band and keeping the band clustered, compared to non-territorial harem stallions. Whenever a territorial stallion does round-up his mares, he directs his activity more toward females low in the female dominance hierarchy than toward high ranking females; the higher ranking females are more involved in mutual grooming with the stallion. Non-territorial stallions treat harem mares more equally (Rubenstein 1978).

what's this work?

17

Social Dominance

Whenever two or more horses are in a group, whether in confinement or free-ranging, they will establish between each other a dominant-subordinate relationship. For the most part, the system of social dominance that develops in groups approximates a linear hierarchy. The number one individual (i.e., alpha) is therefore dominant over all others in the group, the last individual (omega) is submissive to all, and those in middle positions of the hierarchy are in turn dominant to some and submissive to others. Sometimes triangular relationships occur. An example is shown in Table 17.1A where the animal placed sixth in the overall grazing rank order nevertheless showed dominance over the third but not over the other horses in the hierarchy. Occasionally two or more horses appear equal in dominance (note PM and MM in Table 17.1C). Keiper (1976a) found a case of equal dominance in a bachelor band of three young males. Feist and McCullough (1976) found dominant-subordinate relationships clearly evident in bachelor groups, and the alpha male position of each group was overt. Yet they found no consistent hierarchy among individual mares of a harem; dominance among mares appeared dependent upon circumstances. Most other horse researchers have noted some evidence of an established rank order among all social group members, including mares and immatures.

ESTABLISHING AND MAINTAINING RANK

A horse gains a dominant position over another individual by exhibiting enough aggressiveness that the other individual yields or withdraws. The interaction may be violent with kicking, striking, and biting; or it may be mostly pushing with head and neck bumping; it may involve only displays, such as kick threats or bite threats, with no direct physical contact; or avoidance may occur and be almost imperceptible as being an interaction. After one or more interactions and usually within a day or two after first being together in a group, the relationship between any two individuals becomes relatively fixed.

In future encounters, fighting rarely occurs. If the subordinate individual does not yield on its own accord, a threat gesture from the dominant will normally cause withdrawal (Figure 17.1). Both individuals know their status. With the reduction in aggressive interactions, social life becomes more efficient.

In a mixed sex, varied age herd that had been together at least 8 months, Montgomery (1957) noted 488 dominant-submissive interactions. Of these, 74.7% were bites (two-thirds being threats only), 10.3% involved passive avoidance, 8.3% were head bumps, and 6.2% were kicks or kick threats.

Future studies of dominance should be careful to distinguish dominant aggressiveness from submissive defense. Wells and Goldschmidt-Rothschild (1979) found in their study that bite threats were given toward subordinates but kick threats tended to be used as a defensive reaction against dominant animals.

[margin notes: "not like my data", "true →", "while running away", "MORE! DATA."]

Table 17.1: Dominant-Submissive Interactions During Grazing, Drinking, and Leisure in a Group of Six Mares on Pasture

Mares	PR	LTS	PM	MM	CS	CK	Total Threats	Number of Horses Threatened	Apparent Rank Order
				(A) Grazing					
PR		4	11	6	12	7	40	5	PR
LTS	2		5	6	20	9	42	5	LTS
PM				2	7	2	11	3	PM
MM					6	9	15	2	MM
CS	1					11	12	2	CS
CK			4				4	1	CK
				(B) Drinking					
PR				1	1	1	3	3	PR
LTS					4	16	20	2	LTS
PM				1	2	1	4	3	PM
MM						5	5	1	MM
CS						1	1	1	CS
CK			5				5	1	CK
				(C) Leisure					
PR		1	13	9	4	3	30	5	PR
LTS			6	6	14	12	38	4	LTS
PM				5	3	1	9	3	MM↔PM
MM				5		2	2	9	3	
CS							0	0	CS
CK			2				2	1	CK
Body wt, kg	475	475	500	450	430	340			
Age, yr	17	8	18	21	11	3			

Data from McPheeters 1973

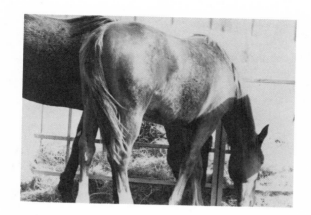

a.

b.

Figure 17.1: Social dominance interaction where a horse high in rank (behind) displaces a subordinate (foreground).

Social dominance can be exhibited in a wide variety of daily activities and is not unique to any sex or age class. Feeding competition is one situation where dominant-submissive relationships can be seen. A dominant will supplant a subordinate at a preferred grazing site or at a feed bucket. Researchers oftentimes provide a source of food to increase interaction frequency so as to more easily determine rank order (e.g., Grzimek 1949d, Tyler 1972, Houpt et al. 1978). Rank order during free grazing does not appear to change when the interaction frequency is increased by providing additional food (Clutton-Brock et al. 1976). Glendinning (1977) found that small groups of orphan foals sequentially followed a rank order when fed with a milk dispensing machine. Dominance relationships among horses can also be seen in such activities as drinking, resting site selection, breeding, sequential rolling and marking behavior as well as during <u>herd movement</u>. argued

Although the general trend remains, a rank order for one type of activity (e.g., feeding) may not be as obvious or exactly the same in another situation (e.g., drinking and leisure in Table 17.1). A horse that is persistent in getting its way while grazing may show far less tenacity for access to water; thus, another horse may make it yield its place at a water trough but would be unable to make it yield its grazing site. Some horses may threaten other horses in primarily one activity, such as grazing, and be passive or submissive in other situations (note CS in Table 17.1). No matter what the activity, threats are not equally distributed among subordinates; most individuals threaten certain horses more than others without any apparent correlation to nearness of rank. Clutton-Brock et al. (1976) found in their field study of Highland ponies that threats occurred throughout the day at an average rate of 1.9 per horse per hour.

Once social dominance is established in a group, the rank order remains quite stable over time (Tyler 1972). Among foals, stable relationships become evident by six months of age. Death or removal of a horse does not cause a change in the dominance relationship of those remaining. Even when a herd is divided, Grzimek (1949d) found the hierarchy in the new groups similar to that observed in the larger herd. Stability seemed greatest at the top and the bottom of the scale. Although some dominance shifts did occur when the herd of 29 yearling colts was divided, Grzimek noted that the average change amounted to only two places up or down the expected order based on the rank in the original large grouping.

FACTORS INFLUENCING RANK

The influence of factors such as age, weight, and height have been studied with regard to dominant-submissive relationships. Age appears to play some role in gaining a social position but is not necessarily decisive (Grzimek 1949d). More often the effect of age in a social group is most evident in the lower part of the hierarchy where the immature members tend to fill the bottom positions (Tyler 1972). Wells and Goldschmidt-Rothschild (1979) found that rank order based on head threats was highly correlated to age in a herd of Camargue horses ($r_s=0.988$, n=25, $P<0.001$). Body size seems to be an important factor in gaining a social position; yet, exceptions can be seen (note CK and PM in Table 17.1). Hechler (1971) found that in the three pony herds he studied, a statistically reliable ranking system could be determined by using weight and height. Tyler (1972) concluded, that among the higher positions in a hierarchy of ponies, size was more important than age in determining social position.

The influence of sex on social dominance has also been investigated. Stebbins (1974) concluded that the interrelationship between sex and age would result in a social rank order as follows: stallion, mare, gelding, male juvenile, female juvenile, male foal, and female foal. In feral herds, an adult stallion typically occupies the alpha position in harem bands as well as in bachelor groups (e.g., Feist 1971, Salter 1978). Jaworowska (1976) found similar results in free-living Polish horses. Berger (1977), however, found that in one of the feral horse bands he observed, the harem stallion was subordinate to the two mares in the group. In non-feral, domesticated herds, it is often observed that males do not necessarily rank above females during maintenance activities. Montgomery (1957) found females occupied intermediate positions among geldings in the herd of eleven he tested during feeding situations. Houpt et al. (1978) found similar variability when mare, stallion, and gelding dyads were tested at a single feed bucket; in their study, they found stallions during the breeding season were dominant over high-ranking mares but submissive to mares low on the mare hierarchy. In the semi-feral herd studied by Wells and Goldschmidt-Rothschild (1979), stallions were subordinate to adult mares with foals in most situations, except when a stallion exhibited <u>driving behavior</u>; the only adult mare without a foal held an intermediate position between the two high-ranking stallions. In social groups containing geldings, sometimes a gelding will occupy the alpha position (and even assume the role of harem stallion) with mares and stallions subordinate to him (e.g., Stebbins 1974).

The length of time a horse has been with a band may affect its position in the dominance hierarchy; yet, it is not well documented in horses. In many other species, newcomers are typically at a disadvantage in an established social group and must work their way up the hierarchy; they typically hold lower positions. Wells and Goldschmidt-Rothschild (1979) postulated that one of the reasons the older stallion in their study was not top ranking was because he was the newcomer. Assertiveness became more and more evident in the stallion as time went along.

Prior experience and physical condition, including endurance, undoubtedly play a role in determining an individual's dominance status. In most cases, harem stallions dominate bachelors; yet, infirm harem stallions eventually yield their position to younger challengers. A band of a crippled stallion observed by Feist (1971) was eventually split into two bands by other stallions. Ebhardt (1957) witnessed a violent fight between an older harem stallion and a less experienced young challenger. Though the young stallion appeared outmatched by the strength and skill of the old stallion and though it repeatedly was in serious danger, it continued to fight. When the endurance of the older stallion finally waned, the young challenger drove off six mares from the harem and the battle ended.

Along with other factors, temperament plays a role in social dominance. Aggressive and persistent horses regardless of weight, height, sex, or length of residence in a band achieve higher ranks than more passive individuals (e.g., Ebhardt 1957, Blakeslee 1974). Tyler (1969) observed numerous cases among New Forest ponies where small, young, but aggressive mares were dominant over larger, older mares. She even saw a very aggressive 5-year-old mare arrive as a stranger in a new area and within two weeks the newcomer was dominant over all other mares. Stebbins (1974) found some mares in her study became more aggressive once their foal was born; thus through that summer those mares showed temporary dominance over, for example, mares without foals and geldings. Boyd (1980) found no evidence of rank order change after foaling in the feral horses she observed. Estrus has not been found to alter a mare's rank; however, a mare who immigrates into a new band during estrus may consort with the harem stallion and therefore because of his proximity initially receive few challenges from others in the band.

It has been suggested that dominance may be related to the learning ability

no data ↓

*is not this
is what you've
been talking
about !*

or intelligence of a horse (e.g., Blackeslee 1974). Nevertheless, Rudman et al. (1980) found no correlation in the experiments they conducted.

INFLUENCE OF RANK ORDER ON DAILY ACTIVITY

On the Pryor Mountain Wild Horse Range, Feist (1971) observed what appeared to be a relationship between dominance and rolling activity. In all instances where some or all members of a group rolled in succession, the dominant stallion was the last to roll. Stebbins (1974) noted some tendency for rolling to occur sequentially through social facilitation, but in her study she found no correlation of rolling to the dominance rank order.

Foals normally rank low in dominance; however, while near their mother they share the mare's dominance rank. For example, the foal of a dominant mare will not be threatened by mares subordinate to the mother provided the foal is close to the mother; yet when the same foal moves several meters away, it is threatened by the subordinate mares (Tyler 1972). Immature horses often exhibit snapping (tooth-clapping) when approached or challenged by adults other than their mother. This submissive gesture is especially obvious as foals approach the dominant stallion.

Dominant stallions tend to urinate or defecate on top of the eliminations of subordinate males. Feist and McCullough (1976) found this strictly adhered to in bachelor groups, but in harem bands some exceptions were seen. In a mixed-sex peer group, consisting of one mare and six males, they noted when the mare or a low-ranking male defecated the rest of the band in ascending order of dominance would in turn add their excrement. Compared to other males, harem stallions show a greater tendency to mark excrement they have seen another horse deposit, especially that of mature mares and especially during the breeding season (Turner et al. 1979, 1981). Stebbins (1974) observed that during encounters between stallions the subordinate male preceded the dominant male when scent marking on the feces of a female. Under feral conditions, dominant stallions from two different bands often interact at large well-established fecal piles (e.g., Feist 1971, Welsh 1973, Salter 1978).

When one band encounters another, a high-ranking representative (commonly the dominant stallion) from each group comes forward to interact. The other members of the two groups usually wait the outcome of the interaction and abide by the outcome without entering the conflict themselves. Under conditions of an exceptionally scarce water supply, Miller and Denniston (1979) observed when one of 16 bands attempted to displace another band from the remaining communal water source, the dominant stallion of the approaching band led the aggression in many instances; yet sometimes a single female, or male-female pair, or a pair of males led the attack. The harem stallion was not necessarily a member of such aggressive pairs.

There appears to be no correlation between interband dominance and the number of adult males in a band; however, there is a direct relationship between band size and the number of bands dominated (Berger 1977, Miller and Denniston 1979). Bachelor males, whether alone or in groups, tend to rank low in interactions with other groups; the status of the same males improves after acquiring mares (Miller and Denniston 1979).

Social characteristics within a group, such as nearest neighbors and mutual grooming, can be influenced by dominance rank. For example, Clutton-Brock et al. (1976) observed in a small herd of Highland ponies that most time is spent close to individuals of similar rank and age. They also observed that the ponies groomed most with individuals of similar rank and age. Pauses in the grooming sessions were initiated more by the higher-ranking partner. The higher-ranking member of a grooming pair was more likely to start grooming in the Highland pony herd. On the other hand, Tyler (1969) found dominant ponies she observed initiated only 38% of mutual grooming bouts when partners were observed to start asynchronously.

During long distance travel, the position individuals take in the moving column is, in part, affected by social dominance. Any individual in a social group may initiate group movement; yet Tyler (1969) noticed that if the initiating individual was a young animal it soon stopped until a more dominant pony had overtaken it. In groups with two or more mares, the more dominant mare proceeds ahead of the other mares, each accompanied by their foal followed in turn by the next oldest offspring. The position the dominant stallion takes appears to be influenced by environmental circumstances. Feist (1971) found stallions were either leading or noticeably driving their band in 76.7% of the band movements observed; the stallion was at the front in 66.7% of the cases. When other bands were nearby or disturbances were present, the stallion took a rear position and herded the band away. In multimale groups observed by Miller (1980), the alpha stallion often led a single file column followed sequentially by the more subordinate males.

Social dominance influences reproduction in various ways. The lowest-ranking males of a herd seldom have an opportunity to breed mature mares during estrus. Usually fertile mares in estrus are tended and mated by a dominant stallion. In free-roaming horses, an adult mare is seldom alone but is part of a harem band; the harem stallion is her consort. Sexual harassment may be less for harem mares than for solitary ones because the proximity of the dominant stallion would intimidate other males. Mares tend to reject subordinate males but occasionally are seen to approach, nuzzle, and even present their genital area to dominant stallions (Salter 1978). Stebbins (1974) saw many instances where dominant mares may influence reproductive success of subordinate mares by chasing young mares away from courting stallions. Stebbins also noticed mares with foals commonly were dominant over mares without foals; yet, whether there is a cause-effect relationship of dominance on pregnancy rate or foaling rate needs further study. Asa et al. (1979) reported stallions in their study usually selected the dominant estrous mare for copulation whenever more than one estrous mare was present.

In feral situations, one harem stallion performs most matings; nevertheless, secondary stallions in a band as well as outside males achieve some matings. Feist and McCullough (1976) found seven out of eight successful copulations were by harem stallions. In single male bands, Miller (1979) also found harem stallions did not necessarily accomplish all matings; males outside the band were observed to do as much as 28% of the matings with harem mares. In harem bands with more than one stallion, Miller noted the dominant male participated in only 54% of the observed matings.

Just before giving birth, mares sometimes separate temporarily from their band. Blakeslee (1974) found the dominant mares she observed tended to spend more time at the birth site and did not rejoin their group as soon as did subordinate mares.

18

Agonistic Behavior

Overt aggression invariably captures the attention and interest of even the most casual observer. One of the most dramatic and awesome spectacles of horse behavior is certainly spirited combat between two stallions. Nevertheless, associated with aggression are a variety of other behavioral patterns, such as alarm, threats, submissive gestures, avoidance, and flight; yet these are often disregarded or overlooked. To the serious observer, however, all of these behaviors have relevance and are encompassed by the term agonistic behavior.

ALERT, ALARM AND FLIGHT

As an initial response to an environment stimulus (e.g., a new object, a sound, or an intruder), a horse becomes alert and attempts to orient the sensory receptors of the head toward the stimulus source. Recurring or minor sounds may cause only an ear to rotate while the horse continues, for example, to rest or forage; yet often, stimuli are of sufficient type and intensity that the horse raises its head and investigates more extensively. Other activities, such as walking or chewing, may cease. The alert posture, consisting of an elevated neck with intently oriented head and ears with nostrils sometimes slightly dilated (Figure 18.1), may induce similar alertness in neighboring horses. Looking, listening, and smelling test the situation. If the source of stimulation is deemed unobtrusive after the initial investigation, the horses may resume their previous activities.

If an alerted horse continues to be stimulated, additional investigative and agonistic behaviors occur. If recumbent, the horse gets onto its feet. The vigilant horse may begin to exhibit restlessness as if concerned about the stimulus. A nicker is sometimes emitted, especially by mares to their foal. Slight elevation of the tail may occur. The body and limbs are readied for potential locomotion. If the stimulus is approaching, withdrawal may occur. Spatial separation

175

between social companions becomes less. If the stimulus is moving away, watchfulness with little locomotion occurs. If the source of stimulation is stationary, the alert yet curious horse may begin moving in that direction with ears and head oriented directly toward the stimulus. When the stimulus is approached with apprehension, the direction of travel is circuitously angled rather than a direct approach. Social companions usually proceed to investigate together as a tightly clustered group, with one or two more dominant individuals in the lead. Feist (1971) found the dominant stallion of feral bands led such investigations; group members were close behind and attentive to his cues.

Figure 18.1: Alertness showing elevated neck and intense orientation of the ears and eyes.

With increased stimulation and arousal, alarm may be exhibited. The eyelids open widely exposing white scleral tissue, the neck elevates fully, the nostrils dilate, and direct motion toward the stimulus ceases. Defecation and nervous pawing may occur. One or two members of a group may then emit an explosive blow of air through the nostrils; the sound may be repeated. The entire group shows full arousal. The lead horse, in a jerky manner, may then step in the general direction of the stimulus as if to entice or challenge while still at a safe distance. The horses nervously stare at the suspicious object. Commonly the horse leading the initiative produces another noisy blow of air as it trots a few steps circuitously in another direction and again stares.

When the source of stimulation remains a threat, withdrawal commences. Horses sometimes withdraw in stages, stopping periodically to stare at the stimulus. If the stimulus remains stationary and thus becomes less obtrusive, a hesitant return to investigate may occur. The flight response and the flight distance depend on the intensity of the situation. In a group of unhandled horses, Zeeb (1963) found standing and walking humans were avoided, keeping a distance of 3-5 meters. When the same horses were confronted with a man on all four limbs, the distance of avoidance increased; yet they approached cautiously to investigate if the man remained motionless. Exaggerated movement by the quadruped-like man would then cause flight to occur. The horses as a group would withdraw at a trot for 100 meters.

As the horse nearest an alarming stimulus pivots away from the stimulus and takes flight, its companions do so also. Foals remain at their mother's side. Sometimes allelomimetic behavior (mutual mimicry) within the group increases the flight response, causing more alarm and greater withdrawal than might otherwise be shown. Allelomimetic behavior may also occur earlier when the group makes its approach and investigates. Perhaps a single horse would not approach as closely as a group.

Once group withdrawal commences, one individual may direct the flight by leading or by driving the group from behind. The horse that initiates the flight may be in front for awhile, but in prolonged flight that position may change (Berger 1975). Feist and McCullough (1976) found the dominant stallion displayed the leadership trait in the feral bands they observed. When flight was because of an intruder, the stallion was usually positioned between the band and the intruder as movement continued.

Flight occurs with a speed, manner, and distance relative to the stimulus and situation. Reactions are swift and reflex-like when surprise occurs; in such cases, alert, alarm, and flight may appear to occur simultaneously. After the initial response, a horse usually appears to regulate its actions so as to meet the needs of the situation and not to be excessive in its flight response. When a horse reacts to an object it has approached closely with its outstretched head, withdrawal may be primarily a quick retraction and elevation of the neck and head, often swinging them to one side. In such a manner, a horse withdraws from a kick or strike attempt by another horse. A sting on the muzzle or sudden bad taste while eating will also cause this response. When startled by something in front or to one side, a horse turns quickly and moves away for a meter or more; rearing may occur if the forelegs are threatened. When threatened from behind, the response is a lunge forward, sometimes with a lowered pelvis. A horse sufficiently startled by a touch to the hindquarters will give such a forward lunge response as well as raise and turn its head to investigate.

In most cases of flight where locomotion occurs, the trot is the gait used in withdrawal. Simple avoidance occurs at a walk. When extremely alarmed or pursued, a horse may use a gallop. Foals use a gallop more readily than grown horses when in flight. In responding to intruders, Feist and McCullough (1976) observed feral bands moved 50 to 100 m before stopping to watch the intruder again. They also noted that if the stallion remained calm the rest of the group, though originally alarmed, would soon calm down. In the feral bands observed by Berger (1975), flight covered a distance of 30 to 110 m when not pursued; nervous mares initiated most flight responses of the bands (14 out of 16). Foals were among the last members of a band to take flight, and nearby bands merely became curious when another band took flight rapidly.

When physically restrained, foals and older horses often use a series of pushing, pulling, and twisting maneuvers in an attempt to struggle free. Repeated bouts may occur, but eventually struggles subside. Horses with prior experience of restraint often yield sooner than naive individuals. Foals a few hours of age as well as older horses can be taught to remain calm under human restraint.

Shyness in horses varies between individuals and appears to be an effect of both heredity and prior experience. Fear responses can be reduced through habituation by repeatedly exposing a horse to inconsequential stimuli. Grzimek (1944a) tried to experimentally demonstrate differences in timidity of horses by measuring delay interval and approach reaction of horses to pictures, other horses, and so on. Under his experimental conditions, he found no appreciable difference between sexes, breeds, or test situation. Individuals 11 years and older showed more reluctance to respond than did experimental subjects 10 or younger.

AGGRESSION

The aggressive displays of horses range from relatively mild, subtle acts to intensely violent displays depending on the circumstances. Not only horses but man and other annoyances can be aggressively challenged (cf. Zeeb 1959a). The aggressive distance of a horse varies with the situation but is commonly one or two meters. Normally horses are conservative and display the minimal amount of aggression the situation requires. Thus, threats are far more common than violent contacts. Berger (1977), for example, found only 24% of the 1162 intra-band aggressive acts he recorded around a water hole involved anything more than a threat or mild push; a similar 22.6% occurrence of violent-type aggression was noted by Baskin (1976) for horses competing for winter pasturage (n=111).

The more mild forms of aggression include the laying back of the ears, lowering and extending the head, shifting the hindquarters toward an opponent, and using the body to block or push the opponent. The first sign of displeasure or aggressive intent is generally the posteriorly directed ears compressed against the skull. The ears-laid-back display accompanies aggressive acts; at the peak of those acts, the ears are most compressed.

When the object causing aggression is in front of a horse, the horse tends to react initially using a head gesture. Whereas, when the object is behind, the horse may instead shift the hindquarters toward the object.

A body block is used by a mare with foal as well as by a harem stallion to screen a foal or harem, respectively, and thus to intimidate an intruder who is nearby. Mild pushing with head, neck, or shoulder occurs to displace an opponent, for example, from a feed bucket.

Mid-level aggression is displayed by threats to bite, strike, and kick as well as extended-head gesturing with sideward swing or up and down motion. Vigorous tail switching and even slight hopping motions with the hindquarters may occur prior to kick threats. Threats to strike or kick sometimes involve head and body bumping between opponents and are often accompanied by harsh vocal squeals. In a strike threat one or both forelegs are lifted off the ground, and in kick threats one or both hindlegs are gestured; yet both displays lack complete effort and are restrained. Saddle horses, when being readied for a ride, sometimes seem to protest by stamping the ground with an abbreviated strike threat. Knocking using a hindfoot is a similar aggressive gesture of protest.

Bite threats are the most common of the sudden mid-level aggressive displays. They may consist of a head swing with slightly opened mouth or consist of a nipping motion toward an opponent using an extended head as well as neck (Figure 18.2). Such gestures are delivered toward annoyances in front or to the side of the aggressor. In most cases, it is apparent that the aggressor is only threatening because no serious effort is made to achieve contact or to actually bite the other horse. Bite-like gestures are often given toward flying insects that have landed on the horse's forelegs, back, barrel, or flanks. Bite threats are often directed toward an opponent's head, shoulder, or chest and occasionally their forelegs. When the aggressor is behind the recipient horse, such as when a stallion is about to drive a herd member, a bite threat is directed at the hindquarters; this causes forward motion in the recipient as it flees.

Figure 18.2: Bite threat. (Photo courtesy of P. Malkas)

Although using the mouth to grip an opponent's mane is common in play fighting (Schoen et al. 1976), seldom is non-play aggression displayed by holding onto an opponent. Nevertheless, Feist and McCullough (1976) reported one instance where an immature male displayed submissive snapping to a harem stallion, but the stallion with ears laid back reached out and bit the immature on the shoulder. The stallion, fixed in this position, held onto the skin of the shoulder for about one minute before releasing the young male.

The low neck, head-extended display of aggression given by an approaching horse generally causes withdrawal and thus displacement of others, such as from food or water. The so-called snaking pattern with extended head and the neck lowered to or below horizontal is used by stallions and dominant mares to herd or drive others. Biting may be feigned. Nodding or swinging the neck accentuates the display.

High-level aggression involves serious efforts to bite, strike, or kick an opponent; physical contact is attempted. Yet, even then, some gradation in effort appears. Sometimes the attacker makes contact but refrains from putting full energy into the onslaught. At other times, the attacker does put forth full effort. Contact can result in wounding the opponent, especially when the aggressor uses extreme effort. An opponent, however, is usually cautious enough to dodge most attacks. Unless both individuals are intent on a fight, direct contact often fails.

When a fight ensues, combatants attempt to subdue their opponent by strategically placed bites and by knocking the opponent off balance. Maneuvers are usually forceful and swift. Hindleg bites and circling sometimes occur (Figure 18.3). At other times, fights emphasize rearing and biting at the opponent's head and neck (Figure 18.4).

Occasionally horses use the forelegs alternately in a brief attack when in contact with an opponent. This boxing-like pattern often occurs when two fighting horses rear on their hindlegs and proceed to make contact with the head and forelegs. The alternate striking pattern with rearing is sometimes used against intruders, such as canids.

Figure 18.3: During rapid circling and hindleg biting one combatant has managed to knock his opponent off balance. (Photo courtesy of P. Malkas)

Figure 18.4: Intensive aggression with rearing and head biting. (Photo courtesy of P. Malkas)

Horses a few hours of age and older are capable of aggressive acts. Unsuspecting human handlers are sometimes seriously injured by a kick of a neonatal foal. Skillfulness in using agonistic behaviors increases with experience as well as through physiological and morphological maturation.

Environmental context, dominance status as well as sex and age affect the type of aggressive display presented. In studying the social relationships of a herd of Camargue horses, Wells and Goldschmidt-Rothschild (1979) found mares gave most aggressive head gestures to their yearling offspring and to stallions but fewest to their foal. Mares gave most kick threats (including aggressively orienting their rump) to their foal and yearling as well as to stallions (especially during copulatory attempts). Stallions, yearlings, and foals used head gestures within their own age group and with younger animals, whereas they tended to use hindquarter threats against individuals dominant to themselves. Foals, but rarely yearlings, directed kick threats to their mother. Kick threats were common in play of non-adults; head threats were not. The general conclusion of the researchers was that head threats are given to subordinates in situations such as grazing, shelter seeking, and maintaining individual distance; kick threats, on the other hand, are especially common in contexts such as play and copulation and tend to be used defensively against dominant individuals.

Aggression can occur between and within any sex or age class. Prolonged aggression, however, seldom occurs among foals, except as play. Adults sometimes seriously threaten young horses. Foals that attempt to approach or suckle mares other than their mother are usually rebuffed by the mare's show of force. Horses harried by others may redirect aggression to nearby subordinates. Stallions can be especially threatening to subordinate horses when in conflict situations during sexual interactions. Fatal maulings can occur. Tyler (1969) witnessed stallion attacks on young mares and foals. On one occasion, a recently released stallion had been driving for over 30 minutes a mare accompanied by her foal. The mare was not in full estrus. Eventually the foal got separated from its mother; the stallion rushed to it, grabbed its neck with his teeth, and shook the foal until the mare intervened.

Fights between adults of the same sex can often be intense. Mares occasionally attack other mares. Dominant mares sometimes become aggressive toward subordinate mares who approach a stallion while in estrus. Furthermore, kick fights lasting 2-3 minutes can occur so as to determine dominance when strange mares challenge each other (e.g., Tyler 1969). Fights between stallions can be highly ritualized and at times may be violent.

INTERACTIONS BETWEEN STALLIONS

Interactions between stallions commonly involve ritualized displays unlike other forms of agonistic behavior in horses. Conflicts between mature, free-ranging stallions function in part as a spacing mechanism, inducing harem stallions and their social unit to remain distinct and separate from other stallions and their bands. Agonistic interactions also provide a mechanism for bachelor males to test the ability of a harem stallion to maintain his status and to retain mares. Ritualization has provided a means to interact and to demonstrate dominance without resorting to violence (Tschanz 1979); therefore, separation often occurs without serious physical combat and severe injury having taken place. Interactions between two harem stallions tend to be less intense than interactions between a harem stallion and a bachelor (Salter 1978).

The ritualized interactions of stallions take place as recurring sequences of the following stages: (1) staring while standing, (2) body posturing and locomotor displays, (3) close olfactory investigation, (4) squeals, forequarter threats, and pushing, and (5) fecal pile displays. Sometimes one stage is omitted or shortened to move on to the next stage. Separation may occur following any phase or after several repetitions of the sequence. Separation after a fecal pile display is most common. Interactions between stallions may last only a few minutes or for more than an hour. Salter (1978) observed one interaction to last nearly 1.5 hours when a harem stallion interacted with a group of young bachelors.

The initial stage, the stare, involves one or both stallions standing and looking toward the other while some distance apart. The alert posture is used with ears forward. Whinnying, tail switching, and pawing sometimes occur. Stallions often study passing bands or those grazing nearby. A slight approach may occur before again standing and watching. Salter (1978) concluded that about half of all harem stallion interactions do not proceed beyond this stage.

If stallions do interact further, they proceed to display their status and intent using visual signals. With neck arched, head tightly flexed, ears forward, and tail elevated, dominant stallions approach each other or move parallel in the same direction. The head and neck may be moved up and down, causing the mane and forelock to be flung about thus accentuating the display. The forelegs are lifted high, and all hooves are forcefully placed on the ground in an exaggerated trot. Sometimes the forelegs are swung forward in the motion of a strike threat. When the stallions are still apart and each displays from the vicinity of separate fecal piles, they may proceed to paw and smell the pile at their location, add their own feces, then again smell the pile. The stallions may continue to interact after the visual display or conclude the discourse and move off in different directions.

When interactions continue, visual displays are followed by a stage of close investigation where each stallion smells and exhales at the nostrils and muzzle of the other. Olfactory investigation may proceed to other parts of the opponent's body, often following a sequence of neck, withers, flank, genitals, and finally the rump and perianal region.

At some point during the olfactory investigation, one or both stallions suddenly squeal and threaten the other with a bite threat or strike threat while slightly rearing. Biting and striking continue to be attempted but are usually blocked by countermaneuvers of bumping and pushing with the head, neck, and shoulder. The forelegs are sometimes used alternately in a bout of striking motions. Biting of the hindlegs may occur causing the opponent to tuck and swing away the hindquarters. Occasionally the hindquarters are used for bumping. A kick attempt with one or both hindlegs as well as the infrequent chase tend to be reserved for the end of a bout of active fighting. In the initial sequence of aggressive exchanges, the body contact phase tends not to be prolonged or as intense as subsequent sequences.

The stallions may then proceed to a nearby communal fecal pile and continue to interact. The fecal piles of feral horses tend to be 1-2 m in diameter yet may be over 7 m (23 ft) in length (Feist and McCullough 1976). In unison or taking turns, the stallions smell the pile, defecate upon it, then turn and smell again. Smelling of the opponent's feces is typical. Feist and McCullough (1976) found no relationship between which stallion defecated first and which either initiated or won the interaction; however, in "mock fights" between immature or bachelor males, the dominant always defecated last. Welsh (1973) concluded that on Sable Island, the defending stallion was first to mark the pile, and the stallion of the band that originally approached and initiated the interaction was the last to defecate.

If the stallions still do not separate after the fecal pile ritual, another sequence begins repeating the various stages or phases of the overall interaction. With each sequence, one phase may be emphasized more than another and the intensity of combat may increase. Additional sequences occur until one or both stallions withdraw with their band. Members of a stallion's band seldom participate in the interactions but remain in the vicinity until the fight is over.

Violent, non-ritualized fights occasionally occur between stallions. Such fights begin by one or both stallions suddenly charging with little or no preliminary posturing. Bites, foreleg strikes, and hindleg kicks are given in earnest. Bloody wounds and broken bones can occur. Soon one individual withdraws and may be chased as well as bitten by the winner. Such fights seem to occur most often when an alien male is rapidly approaching a band of another male or is suddenly found harassing a mare of the band.

Most aggressive interactions between stallions are the result of bands coming too close together or because of disputes over male status and conflicts over mares. Of the 83 aggressive interactions Feist (1971) observed which involved at least one stallion (harem stallions, lone bachelors, or bachelor group), 37 seemed to occur to maintain linear spacing between groups or a lone bachelor, 18 seemed to be the result of males challenging the position of harem males, 12 were associated with attempts to steal a mare, 10 occurred within bachelor groups as dominance order interactions or "mock fights," 4 occurred when a mare separated from her band was being retrieved, 1 was a "mock fight" between a harem stallion and male foal, and 1 was a complicated battle involving 5 harem stallions and 2 immature males.

SUBMISSION

Many instances of submission in horses occur without attracting much attention from an observer. For example, when a subordinate is approached by a dominant individual, the subordinate often seems to saunter away on its own initiative before the dominant is close by. Thus further interactions are avoided. Once the dominant individual moves away, the subordinate then returns to resume its previous drinking, foraging, or other activity. When a subordinate does not react as early to the approach of a dominant, its avoidance response will show deliberate, somewhat hasty withdrawal movements, especially when the dominant is within 2 m; its ears are laid back as it steps out of the way. Thus, the common form of submission is to move away.

Upon yielding to a dominant or aggressor, young horses often move closer to their mother or peer companion, as if for comfort. When a foal returns to its mother after being threatened, it commonly initiates a bout of nursing (Tyler 1969, Blakeslee 1974).

Moving away from an aggressor is not always possible; yet submissiveness can still be shown. A horse that is trying to avoid an attack by a threatening individual close at hand will toss its head and neck upward and often to one side if the aggressor is in front, or if the aggressor is behind, it will tuck its tail and flex the hindlegs while attempting to shift its rump away from the aggressor. In the head toss, the neck is raised maximally and the head is either tightly flexed or elevated close to or above horizontal. The eyelids open widely as the horse stares at the aggressor, exposing the light colored scleral tissues around the iris. The nictitating membrane oftentimes covers the anterior portion of the eye as the head is elevated. The submissive or fearful head response occurs to rough human handling when the forequarters are abused, for example, with a whip, lead chain, or bit. The tightly tucked tail and crouching of the hindquarters occurs when a horse is hit along the back, rump, or hindlegs (Dark 1975).

Young horses three years of age or younger show a specialized form of submissiveness that has been called "snapping" (Tyler 1969), "teeth-clapping" (Feist 1971), "jaw-waving" (Blakeslee 1974), and "Unterlegenheitsgebärde" (Zeeb 1959b), and so on. An immature horse initiates the display by extending the head and opening the mouth, usually with the corners drawn back (Figure 18.5). The ears may splay laterally. The individual then begins a series of jaw

Figure 18.5: Snapping display of submissive immature horses toward adult stallions. [Photos courtesy of R.R. Keiper (a,b) and P. Malkas (c)]

motions, opening and partially closing the mouth without the lips (nor often-times the teeth) making contact. In some cases a slight sucking sound occurs with the jaw pattern as the tongue is drawn against the roof of the mouth (Schäfer 1975).

The snapping display of immature horses is given especially when the indi-vidual seems apprehensive about the proximity of a larger or more dominant animal. Tyler (1969) observed snapping in neonates when mares first turned to their foal after birth. Williams (1974) studied the displays in several orphan and mother-reared foals. One of the foals raised with its mother displayed snapping to an approaching cow when one week old. Some foals gave the dis-play only to more mature horses; some responded upon the approach of hu-mans. Machine-reared foals responded to strange humans but not to familiar ones.

Most observers have noted that snapping decreases with age. Of the 252 in-stances reported by Tyler (1969), 58.3% were given by foals, 32.1% by year-lings, 5.2% by 2-year-olds, and 4.4% by 3-year-olds. It appears that as young-sters gain more experience they learn to discriminate threatening from non-threatening situations. Subsequently, snapping tends to be limited to potentially threatening conspecifics when other agonistic responses are not prudent.

Foals and yearlings commonly display snapping when approached directly by adult mares and stallions. Foals searching for their mother often exhibit snapping as they approach each horse. Tyler (1969) noticed searching young foals even gave the display as they approached their own mother until they recognized her. Tyler also observed that young mares in their first estrus showed snapping when stallions sniffed, licked, or nibbled them as well as dur-ing copulation. When mares with foals are investigated by stallions, the foal commonly gives the jaw movement display; the mare rarely does.

Stallions induce more snapping displays than do other horses. Wells and Goldschmidt-Rothschild (1979) found that among yearlings and foals the males gave the display more often than females in the same age class. Snapping toward stallions seemed to be induced by mere proximity, but the display given to mares was more often in response to a direct threat from the mare.

Feist and McCullough (1976) noted several instances where snapping toward the dominant stallion was given by an immature male immediately fol-lowing an interaction between dominant stallions of different bands, after mare tending, and after the dominant had been absent from the band for awhile.

Zeeb (1959b) suggested that snapping may have originated from social grooming and that the display of jaw movements could be intention move-ments. Feist and McCullough (1976) pointed out that grooming another indi-vidual can in some instances initially be an appeasement activity and that both snapping and appeasement grooming may be a ritualization which allows an immature to express subordination and perhaps avoid aggression. Aggression seldom occurs; yet because the immature horse does not necessarily withdraw and defuse the threatening circumstance, the situation remains tenuous until the mature horse accepts the gesture of submission. If the mature horse does not accept the closeness of the immature, it will show further aggression.

19

Communicative Behavior

Throughout much of each day, horses emit signals that convey information. The information may pertain to the horse's intentions, present activity, social status, mood, identity, physiological condition, or perhaps its awareness or concern about something in the surroundings. Often the horse itself may not be aware it is emitting such signals; yet when another animal perceives and interprets any of the messages, information is exchanged and communication is *def ?* achieved. The receiver may then emit signals relevant to the information it has just obtained and thus establish two-way communication. The information exchange is typically between horses; however, communication can also occur with other species, such as with humans. Communicative exchanges are fundamental to horse handling as well as to social interactions and group living.

Communicative signals between horses can be visual, acoustical, tactile, or chemical. Interactions often involve more than one mode. The function attributed to the signals are usually inferences made from the environmental context and the reactions of the sender and recipient. Gradations of many expressive patterns occur depending upon the degree of stimulation and situation.

VISUAL EXPRESSIONS

Various parts of the body are used in visual expressions; their effectiveness for communication may be as a whole (i.e., collectively), as subunits (e.g., head gestures only), or individually (e.g., only the mouth). Head and leg gestures are common and are accompanied by changes in the position of the ears, tail, and neck plus facial characteristics (Dark 1975; Schäfer 1975, 1978). The long hair of the forelock, mane, tail, and fetlock accentuates visual displays. Visual displays range from exaggerated actions and patterns to subtle expressions, often over-*?* looked by human observers. That horses can perceive and utilize even slight visual cues was clearly demonstrated in the oft cited case of Clever Hans whose intellectual skills were dependent upon being able to read the answers in the gestures of human bystanders (cf. Pfungst 1907).

Stance and overall body posture, without requiring further details of facial features or movement, are useful signals to interpret such things as a horse's mood or physiological condition. The placement of the legs as well as the attitude of the head, neck, and tail provide a comprehensive signal. A horse in prolonged pain, for example, is usually recognized by overall body posture; the weight distribution on the legs may be noticeably shifted and a droopy appearance of the head, neck, and tail occurs.

Leg gestures are common signals in equine social interactions. For example, motions to strike with a foreleg or to kick with a hindfoot are common expressions in agonistic situations. A leg may be raised suddenly into a potential attack position and held momentarily. Oftentimes a kick or strike movement is made thrusting a leg or a pair of legs into the air; in most instances, the legs are either restrained or the aim indirect so as not to achieve contact with the stimulus object. The function of such leg gestures appears to be offensive or defensive warnings to cause withdrawal by the recipient and thus some spatial separation.

Knocking and stamping also occur in agonistic situations. The forceful contact of the legs with the ground adds an auditory component to the gesture. When an individual is eating and seems to object to being crowded by others, it may knock with a hindleg without ceasing its eating activity and, thus, efficiently signal its protest. Ödberg (1973) reported a mare knocked repeatedly with a hindfoot when crowded at a brush pile while eating. In a similar manner, a gelding consuming oats in a field threatened away approaching calves. Ödberg's conclusion was that such knocking stems from an intention movement of a kick.

When a horse with ears laid back aggressively thumps the ground with either a fore- or hindfoot while being prepared for a ride, the gesture seems to signal the horse's objection or protest. Knocking as well as hindleg lift can be frequent when insects and other irritation occur at the belly and flanks; occasionally the hindfoot is used to bump the abdomen under these circumstances. These gestures can serve as warnings to a foal or veterinarian causing the irritation.

Pawing often serves as a visual signal. The movements can vary from slight to exaggerated and can be accentuated by the sound of contacting the substrate. Ödberg (1973) noted that pawing as a possible displacement activity occurs in conflict situations, such as (1) when a horse is aware of food but is unable to reach it, (2) sometimes during eating while in the presence of onlookers, (3) when anticipating locomotion, such as a race or release from restraint, and (4) occasionally by stallions who are delayed while being led to a mare for breeding. Maday (1912) pointed out that pawing often serves to signal a want or need. Pawing can function to scrape, uncover, or test something and at the same time so inform others; it often occurs prior to rolling, during investigation of objects on the ground, and to dig for food or water. Occasionally pawing-like motions signal discomfort, such as during parturition and when the mare struggles to discharge the fetal membranes. Pawing can become conditioned as seen in some begging horses as well as in performing horses (such as circus horses) where reinforcement has strengthened the behavior.

does expressive = comm'tive

Additional leg movements can have signal value. The alternate lifting of the forelegs occurs in restlessness and may become established as the habit called weaving. The treading in place or marking time occasionally shown by *comm* standing horses while being restrained by a rider seems to express the horses desire to move forward. To the rider the movement is more a tactile expression, but to an onlooker it is a visual display. In highly schooled horses, this behavior pattern is developed as a piaffe executed in place.

Locomotor activity can be expressive. For example, brief bouts of trotting using audible hoof contact with the ground and shifts in direction of travel are characteristic of a nervous horse investigating a suspicious object. As the individual withdraws suddenly in alarm, companion horses commonly react to the sudden flight and likewise withdraw. Conversely, slow locomotion in a relaxed manner signals that there is no alarm, that calm prevails.

Other visual cues, such as facial and tail gestures, accompany locomotor activity and undoubtedly clarify most situations. This becomes obvious when observing the approach of a stallion to a potentially receptive mare. The high neck, tightly flexed head, attentive focus of the ears, and elevated tail nearly overshadow the springy prance used by the stallion in locomotion.

Vertical head motions in the form of nodding often occur during approach situations. Stallions oftentimes nod as they approach a mare. Foals nod on some occasions when eagerly approaching their mother.

The lateral motion of the head and neck during weaving is an overt visual display. Since weaving develops as a stereotyped pattern primarily under confinement, the value of the display as a visual signal between horses may be nil. Yet to a person familiar with horse behavior, weaving can be a signal that all is not well with the housing or other management conditions.

Facial expressions vary primarily by changes in the orientation of the ears and eyes; changes in the position of the lips, jaw, and eyelids; changes in shape of the nostrils; and contour changes of the skin surface, especially at the corners of the mouth as well as around the eyes and nostrils. Some features are situation specific; for example, nostril dilation is generally associated with deep breathing and sniffing. Lack of tonus in facial musculature upon death (Figure 19.1) produces an expressionless appearance in contrast to the various displays occurring while alive.

Figure 19.1: Facial characteristics of a dead horse for comparison to expressive displays of living horses. (Dark 1975)

To review the multitudinous head and facial features of potential value in visual communication, it is helpful to group the expressions into sets of related expressions. The scheme my students and I have adopted includes expressions of drowsiness and sleep, forward attention, lateral attention, backward attention, alarm, aggression, sensual pleasure, juvenile snapping, flehmen, and yawn (Dark, 1975, Waring and Dark 1978). Expressions used in the flehmen response and while yawning have been discussed in Chapter 3 and will not be repeated here.

Facial expressions of drowsiness and sleep vary primarily in the amount of eye closure and droop of the lower lip (Figure 19.2). When a horse goes from alert wakefulness to drowsiness, the slight eye closure and relaxed body posture help signal the change. The ears continue to rotate toward surrounding sounds, but other movement is nil.

As sleep progresses, the neck continues to relax and approaches horizontal. The ears relax to a lateral position and cease movement. The eyes close. And in some horses, the lower lip droops noticeably, separating and extending beyond the upper lip. The horse is often standing or in sternal recumbency. In lateral recumbency, eye closure and complete relaxation is typical. Slow-wave sleep may progress to a bout of paradoxical sleep where twitching and movement of legs and facial features occur. As one horse becomes drowsy and sleeps, other members of its social unit appear induced through social facilitation to also relax and sleep.

Figure 19.2: Expressions of drowsiness and sleep. In some horses, the lower lip droops as sleep progresses. (Waring and Dark 1978)

Expressions of forward attention are characterized by anteriorly directed orientation for reception of visual, auditory, olfactory and sometimes tactile cues (Figure 19.3). The ears are up and rotated forward. The eyes are directed forward and appear to emphasize the binocular visual field. The neck and head angle adjusts to facilitate use of the sensory receptors. An elevated neck with head flexion is used for distant visual inspections, whereas head and neck extension occur in close olfactory and tactile investigations. The nostrils are moderately dilated especially when sniffing. The mouth is usually closed.

Figure 19.3: Expressions of forward attention. (Waring and Dark 1978)

0 and 0-5.	Horse standing or moving in alert manner.
0-1-2-3-4.	Investigation or manipulation of material on or near the ground.
4-3-2-1-0.	Ceasing of feeding to observe something in the surroundings.
0-5-6-1 and 0-1.	Inspection, such as naso-nasal greeting or when handed food.
0-5-6-7.	Sexually aroused stallion; horse activated while on a halter line.
0-8-9-10 and 0-8-11-12.	Displayed during energetic locomotion (tail often elevated).
8-9-10.	Horse actively avoiding an object on or near the ground; horse yielding to bit pressure.
0-8-11-12.	Horse approaching a jump.

Expressions of lateral attention are characterized by general relaxation, with the eyes and usually the ears oriented to the side (Figure 19.4). Sensory receptors if not attentive to something along one side may not be focused on anything in particular. Often the horse appears to have no immediate concerns.

The individual may be inactive or in motion; horses relaxed while being ridden or routinely handled may also show these expressions.

Figure 19.4: Expressions of lateral attention. (Waring and Dark 1978)

0-1-2-1.	Horse in relaxed walk; pattern occurs also when shaking.
0-2-3-4-5.	Lowering of head during quiescent grazing; reverse sequence occurs during pauses.
0-6-7-8.	Play fighting, often while facing opponent and achieving periodic head and neck contact.
0-6-9.	Tractable horse at ease with rider.

Expressions of backward attention are characterized by the eyes and usually the ears being rotated to enhance posterior reception (Figure 19.5). The mouth is normally closed unless the horse is vocalizing or has a bit or other object in its mouth. The expressions occur not only with posterior visual and auditory investigation while the head is directed anteriorly, but also the expressions may occur when a horse is stressed, uncomfortable, or seems apprehensive about the rider.

Expressions of alarm show as widely opened eyes, twitching ears, tense mouth, and dilated nostrils (Figure 19.6). Tension throughout the body occurs and is often accompanied by sudden jerky withdrawal movements, cringing, sweating, as well as increased respiration and heart rate. Gradations vary from alert suspicion to extreme fright.

Figure 19.5: Expressions of backward attention. (Waring and Dark 1978)

6-5-4-3-2-1-0.	Sequence shown by a grazing horse that is approached from behind. Reverse sequence occurs with continued vigilence as grazing progresses.
0-1-2-3.	Horse pushing against a restrictive barrier.
3-4-5.	Horse physically exhausted or in discomfort.
4-5.	Facing downwind during a severe storm.
0-7-8-9.	Tugging at bit and reins to succeed in release of rein tension from rider.
8-10-11.	Occurs during strong tension on reins when tack limits head elevation.
0-7-12-13-14.	Variations during head tossing, balking, or bolting often in response to harsh handling by rider.

In expressions of aggression the ears are laid back and compressed against the skull (Figure 19.7). The eyes are alert, open, and generally oriented toward the object causing the aggression. The nostrils are usually dilated and drawn back with wrinkles occurring along the upper, posterior edge. General muscle tension of the body is evident, and the mouth may be open. In extreme cases the incisors may be exposed conspicuously to bite or to threaten biting. When exhibiting a bite, bite threat, or the snaking display the neck is lowered and the head extended. The aggressive expressions vary in intensity and occur in aggressive conflicts between horses and nearby animals.

Figure 19.6: Expressions of alarm. (Waring and Dark 1978)

0-1-2-3.   Frightened horse in locomotion.

0-1-2-3-4. Horse being subjected to roughness by rider when restrictive equipment suppresses head extension and elevation.

0-1-5-6-7. Alarming stimulus beside or below horse.

0-8.   Horse approached by suspicious object.

↘ from scratching of withers

When a horse (either by itself or with the help of others) is rubbed, scratched, or groomed it oftentimes exhibits behavioral evidence that intense pleasure is occurring. The expressions of sensual pleasure are characterized by the extension and action of the upper lip (Figure 19.8). The eyes orient laterally and may close slightly; usually the ears are up. As the tactile sensations continue the upper lip extends more and more and twitches rapidly. If the upper lip contacts something, the object is rubbed using the quivering lip. The nostrils do not dilate but shake in conjunction with the active upper lip. The head extends somewhat and may turn to one side. Heavy breathing, groans, and leaning toward the stimulation may also occur.

In the expression of snapping the head extends, the mouth opens slightly, and the corners of the mouth are drawn back (Figure 19.9). Vertical jaw movements occur, causing a series of chewing-like movements. Mouth closure is usually incomplete during the bouts of jaw movement. Clicking of the teeth or sucking sounds occasionally occur. The lips remain separated; in some cases, only the lower incisors remain visible. Young horses display this gesture toward adults when apprehensive about the approach or possible reaction of the nearby horse.

The display can also occur as a similar response to the nearness of a human, cow, or other large organism. The ears vary in position but tend to spread laterally; the eyes usually orient toward the stimulus source. Submissiveness appears to be expressed.

Figure 19.7: Expressions of aggression. (Waring and Dark 1978)

0-1-2-3-4.	Expression during biting and bite threats.
0-1-2-3-5.	Horse driving or dispersing others, often while swinging head in snake-like manner.
0-6-7-8-9.	Vigorous approach of stallion toward another male.
0-6-10-14.	Pattern shown during aggression with rider, during bucking, and in male-male fighting.
0-10-11.	Expression during kicking and kick threats.
0-10-11-12-13.	Displays occurring with foreleg striking, rearing, pushing, and avoidance. Occurs when handler strikes head of horse with quirt or whip.

Tail displays (Figure 19.10) commonly accompany facial, neck, and leg expressive movements. Furthermore, the tail can independently have signal value, such as in the display of estrus, where the tail is erected and often held to one side. A relaxed horse carries its tail down. The tail is compressed against the hindquarters when a horse is facing downwind during a severe storm, in extreme submission, and when withdrawing with intense fear or alarm. As locomotor movements increase in speed and stride the tail elevates correspondingly to the level of the back or higher, usually with a slight arch. Under exuberant and animated locomotion, the fleshy portion of the tail often reaches vertical; the long hairs of the tail stream behind in a showy display. Kiley-Worthington (1976) concluded that the tail is raised as an intention movement to move faster and lowered as an intention movement to decelerate.

is this reflex resp? which facilitates mutual grooming?

Figure 19.8: Expressions of sensual pleasure. The upper lip extends and twitches. On some occasions, the head may turn. (Waring and Dark 1978)

Figure 19.9: Expression of submission shown by immature horse giving the snapping (Unterlegenheitsgebärde) display. (Waring and Dark 1978)

Figure 19.10: Tail postures and displays. (Waring and Dark 1978)

0.   Relaxed position while standing.

1-2-3-4-5.   Variations progressing from a leisure walk to faster gaits, including jumping while at ease.

0-1-2-6.   Sequence prior to defecation.

1-2-6-7.   Display typical of mare in estrus as well as during urination and copulation.

1-2-6-8-9.   Tail switching at insects and prior to kicking, striking, bucking, and balking. Some lashing in the vertical plane may occur in aggressive displays.

1-2-6-10-   Display during intense exuberance or excitement, usually ac-
11-12.   companied by snorting or blowing and energetic trotting or galloping.

2-6-10-11.   Tail display of stallion during mounting and copulation.

0-13-14-   Display of aggression, alarm, or when horse is not at ease with a
15.   handler.

0-16-17.   Display of extreme fear or submission, prolonged pain, or while facing downwind in severe weather.

During aggression, the fleshy portion of the tail appears to stiffen, thus the distal end extends the tail display posteriorly even with slight elevation. Flying insects are brushed off the hindquarters using tail switching from side to side. Forceful sideward motions of the tail, occasionally with vertical lashing, are commonly shown when a horse is annoyed, such as preparatory to kicking, striking, bucking, and balking. During copulation, the stallion's tail is elevated and rhythmically flexed in the vertical plane during ejaculation. Prior to defecation the tail is elevated. In the mare, the tail is raised and held to one side during urination and copulation. When a stallion is marking using bursts of urine, his tail is elevated more than during normal urination.

Although head, locomotor, and tail displays are used as expressions of sexual interest or activity, there are additional visual sexual signals that need mention. Frequent winking of the vulva by the mare in estrus repeatedly exposes non-pigmented membranes as the clitoris is everted. In conjunction with the elevated tail display, winking may help signal to stallions the mare's level of receptivity. In like manner, the erection of the stallion's penis helps signal sexual readiness of the male.

Besides leg and tail movements, additional visual patterns that occur in response to insect pests and skin irritation are shaking and skin twitching. When irritation is around the face, ears, or neck, head shaking occurs intermittently. Localized twitching of the skin surface occurs, especially with irritation around the shoulders and forelegs. Shaking of the whole body occurs commonly after rolling and, for example, after a saddle is removed. Whether such visible displays are communicative to other horses is not known.

ACOUSTICAL EXPRESSIONS

A variety of sounds are produced by horses. Voiced emissions using the larynx include squeals, nickers, whinnies, and groans. Non-voiced sounds include snorts, blows, snores, hoof-substrate sounds, and incidental sounds from tail switching, eating, grooming, shaking, snapping, coughing, flattus, and male sheath movements. Communicative use is made of voiced sounds and of at least some of the non-voiced emissions. Since spectrographic patterns of horse sounds are numerous, I will use for illustration additional sound spectrograms not used in Waring et al. (1975) and Klingel (1977).

Squeal

Squeals are high-pitched outcrys that show distinct spectrographic appearance of having harmonic quality; the fundamental frequency is usually close to 1 kHz (Figure 19.11a,b). Although higher frequencies are present, most of the sound energy occurs below 4 kHz. In some horses, these sounds seem more harsh than those of others. Squeals are given as single utterances in agonistic situations, apparently as a defensive warning or threat that the annoyed individual will become more reactive if further provoked. Squeals are typical during aggressive interactions between horses, during sexual encounters when the mare protests the stallion's advances, and when a pre- or early-lactating mare objects to being touched anywhere near her obviously sore mammary glands.

As a squeal begins, the mouth is closed, but as the sound continues the corners of the mouth may begin to retract. Mouth opening is not typical but may occur (cf. Kiley 1972, Stevenson 1975). Head extension or flexion as well as lateral head movements may accompany the sound. Depending upon the situation, loudness varies from weak squeaks audible only up to a few meters away to loud screams audible at a hundred meters or more. A mare's protest of a nursing attempt by her neonate is normally less audible than the same mare's squeal response to a teasing stallion.

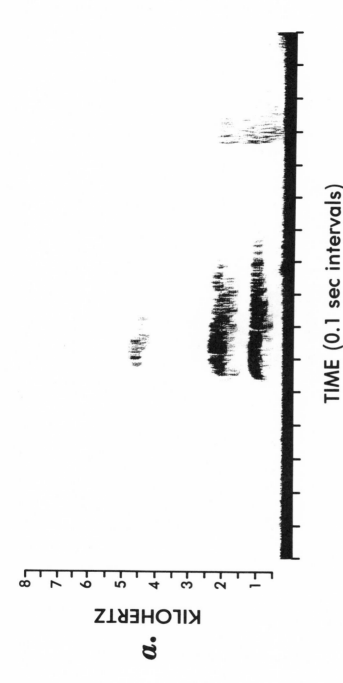

Figure 19.11: Sound spectrograms of common horse sounds: (a,b) squeal, (c) nicker (horse awaiting food), (d) nicker (stallion courting), (e) nicker (mare to foal), (f,g) whinny, (h) groan, (i) blow (alarm), (j) blow (after sniffing), (k) snort, (l) snore (dyspnea). Analyzing filter bandwidth is 300 Hz. (Waring 1971)

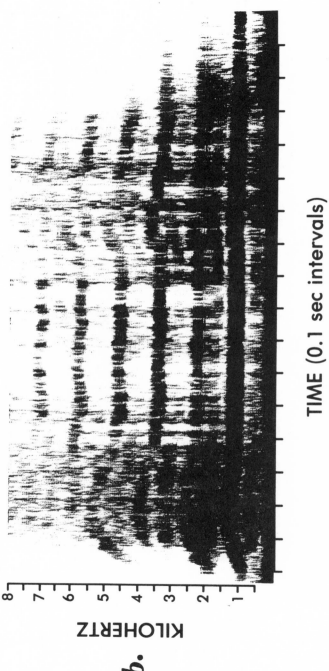

TIME (0.1 sec intervals)

KILOHERTZ

b.

Figure 19.11: (continued)

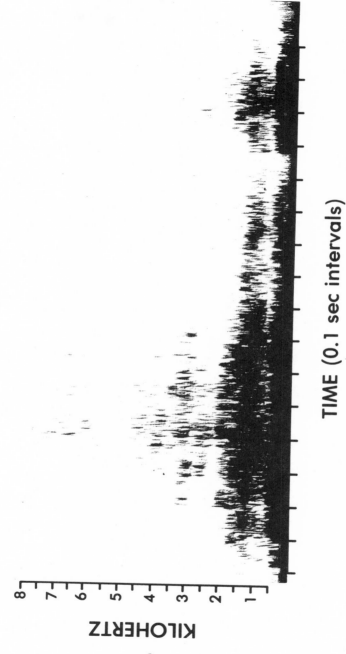

c.

KILOHERTZ

TIME (0.1 sec intervals)

Figure 19.11: (continued)

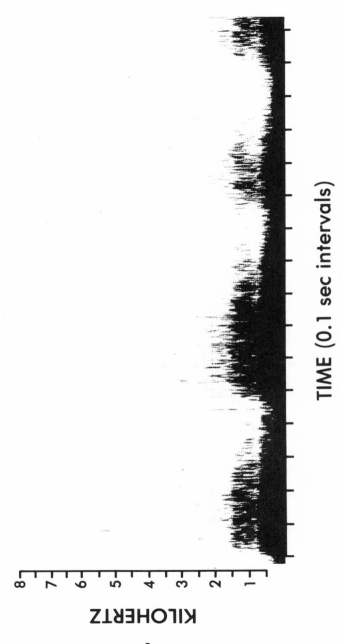

d.

KILOHERTZ

TIME (0.1 sec intervals)

Figure 19.11: (continued)

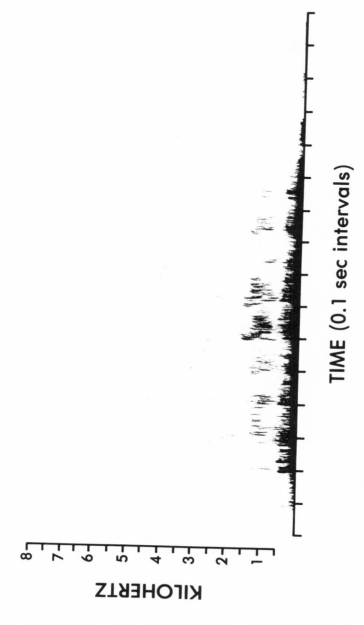

e.

KILOHERTZ

TIME (0.1 sec intervals)

Figure 19.11: (continued)

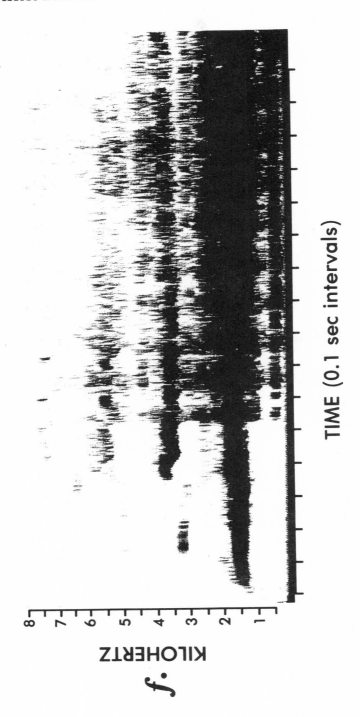

f.

KILOHERTZ

TIME (0.1 sec intervals)

Figure 19.11: (continued)

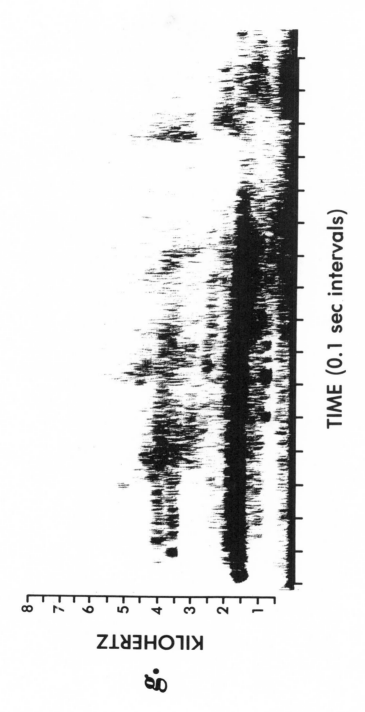

TIME (0.1 sec intervals)

KILOHERTZ

Figure 19.11: (continued)

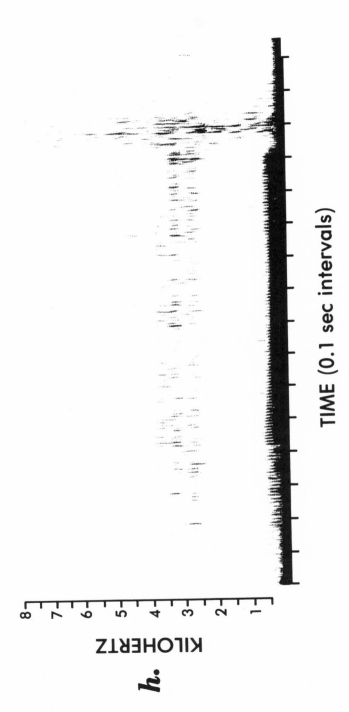

h.

KILOHERTZ

TIME (0.1 sec intervals)

Figure 19.11: (continued)

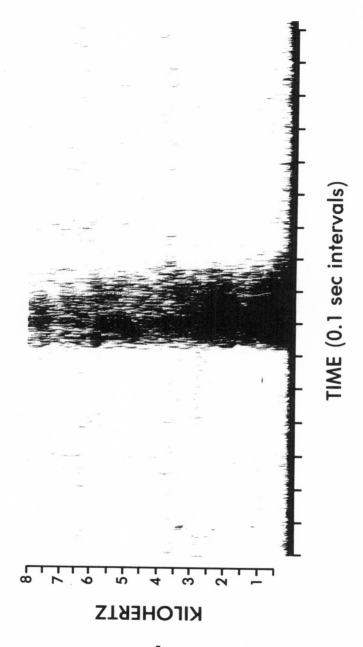

TIME (0.1 sec intervals)

KILOHERTZ

Figure 19.11: (continued)

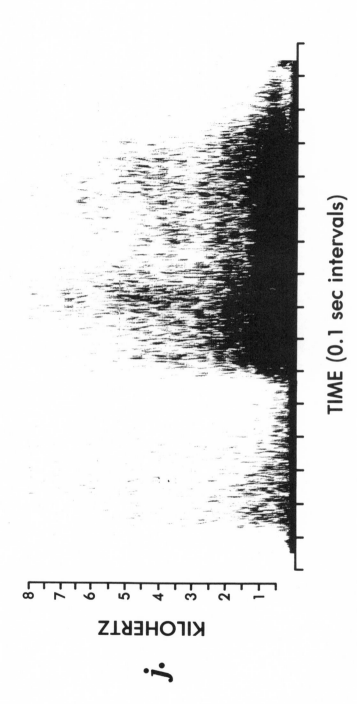

j.

KILOHERTZ

TIME (0.1 sec intervals)

Figure 19.11: (continued)

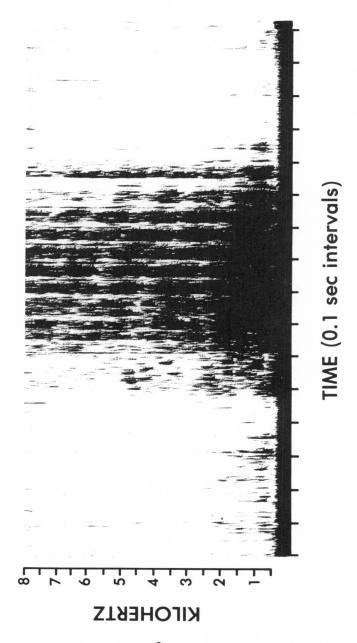

k.

KILOHERTZ

TIME (0.1 sec intervals)

Figure 19.11: (continued)

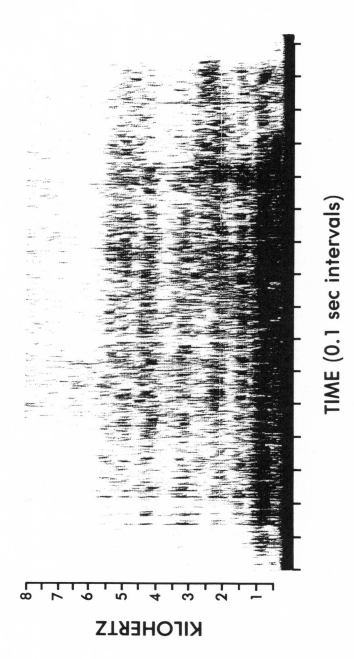

l.

Figure 19.11: (continued)

The duration of squeals varies considerably, ranging from less than 0.1 second to over 1.7 seconds (Table 19.1). Mild protests are the shortest.

Table 19.1: Duration of Sounds Emitted by Horses

	n	Range (msec)	Arithmetic Mean (msec)	Geometric Mean (msec)	Standard Deviation
Squeal	34 (7)*	80–1,720	870	760	340
Nicker	110 (8)	250–1,720	870	780	370
Whinny	56 (13)	500–3,180	1,500	1,410	530
Groan	22 (4)	60–1,690	450	340	380
Blow					
Alarm	20 (6)	210–1,190	470	420	270
Exhalation after sniffing	12 (6)	650–1,330	910	890	210
Snort	25 (5)	280–1,680	900	810	410
Snore					
Pre-blow inhalation	7 (3)	340–460	390	390	50
Dyspnea inhalation	5 (1)	1,040–1,750	1,380	1,350	270

*Number of horses (American Saddlebreds) providing sample.

Data from Waring 1971

Nickers

Three types of nickers have been distinguished. Each are low-pitched, broad-band vocalizations with a gutteral pulsated quality audible to an observer. Resonance bands commonly appear on sound spectrograms. For most nickers, sound energy is typically below 2 kHz; the duration of the nickers I analyzed ranged from 0.2 to 1.7 seconds (Table 19.1).

The nicker commonly heard by most horsemen occurs most often just prior to being fed, i.e., while begging (Figure 19.11c). This type of nicker, whether to man or another horse, announces the horse's presence and anticipation. Easily audible at 30 m, this type of nicker appears least broken into syllables (both to the human ear and on sound spectrograms) than the other two types of nickers. Stevenson (1975) noted the head was raised about 2 cm during the anticipatory nicker, the nostrils were relaxed, the mouth remained closed, and often ear movements occurred.

The second type of nicker (Figure 19.11d) is emitted by stallions during sexual behavior, especially while being led toward a potentially receptive mare. The sounds appear to signal the stallion's sexual interest and are audible at 30 m or more. Repetitious broad-band notes are given, with each stallion having his own individual characteristics, such as in pulsation rate. Repeated head nodding may occur as the stallion maintains a generally collected appearance with head flexed and neck elevated. During the sound the mouth remains closed and the nostrils well open.

The third type of nicker (Figure 19.11e) is typically given by mares to their young foal when potential danger appears or the mare is otherwise concerned about the foal. The low-pitched vocalization, given with mouth closed, expresses the mare's concern and induces the foal to move closer to her side. When iso-lated from the mother, I found neonates can be induced to more readily follow a human if these nickers are imitated and often repeated. A multi-beat, repetitious quality of the mother-foal nicker is typical, e.g., with loudness reaching a peak every 0.1-0.15 second during the duration of the call. Nevertheless, the level of loudness is so low the sound is usually not noticeable beyond the immediate vicinity of the mare and her foal.

Whinny (Neigh)

Whinnies are vocalizations that appear to begin as squeal-like emissions with harmonic structure visible on sound spectrograms and terminate as broad-band patterns similar to nickers (Figure 19.11f,g). Pitch is initially high and appears to drop when the lower-frequency nicker-like portion begins. Whinnies are the longest and most audible of horse sounds, lasting an average of 1.5 seconds (Table 19.1) and often detectable at a distance of 1 km.

Stevenson (1975) noted horses often commenced eye blinking and a head turn just before momentarily elevating the muzzle and emitting a whinny. The nostrils dilated slightly, and at first the mouth was closed. By the time the nicker-like phase of the whinny began, the mouth was open, the corners of the mouth were drawn back, yet the teeth remained covered by the lips. Some walking movements often occurred. During a whinny, the ears and eyes usually exhibit forward attention (cf. Trumler 1959, Schäfer 1975). As the sound ends, the mouth closes and the head returns to a normal position. Oftentimes, the ears then move back and forth alternately. The nostrils remain some-what dilated until the horse further relaxes.

Individual recognition may prove to be one of the additional functions of whinnies, nickers, and possibly other horse sounds. Wolski et al. (1980) found mares tended to whinny more often to playbacks of whinnies of their own foal than to alien foal whinnies, but the difference was not statistically significant (sign test, $P>0.05$). Munaretto (1980) in a similar test found a mare responded vocally to her own foal's vocalization significantly more often than she responded to that of an alien foal ($x^2=19.44$, df=1, $P=0.05$); reciprocally, the foal demon-strated some ability to differentiate between the real and alien mare sounds, but the difference was not significant ($x^2=3.44$, df=1, $P=0.05$). Tyler (1972) ob-served instances where foals responded vocally only to their own mother's whinnies and where band members only replied to the whinnies of lost mem-bers of the same band. Because of a foal's direct orientation and return to its mother upon her nicker, Tyler concluded that by the time foals reach 2 to 3 weeks of age they can identify their own mother's nicker. Further study on in-dividual acoustical recognition is warranted.

When horses become separated, such as a mare and foal or peer companions, one or both individuals often whinny to maintain or to regain contact. At other times, whinnies occur when horses seem inquisitive after seeing a horse in the distance or when they become curious about certain familiar sounds occurring out of view. Whinnies, therefore, seem to facilitate social contact while at a dis-tance. Under playback situations, both whinnies and nickers elicit more atten-tive responses from test horses than do squeals (Dixon 1967, Ödberg 1969).

Groan

Groans are monotone vocalizations that to the human ear appear non-pulsated. Yet these hum-like sounds under 300 Hz bandwidth analysis may show very rapid pulsation as well as a resonance band on sound spectrograms (Figure 19.11h). The voiced groan may be followed immediately by a broad-band, non-voiced but audible completion of the same exhalation. The duration of groans varies from approximately 0.1 second to 1.7 seconds (Table 19.1). The sounds often seem to be expressions of prolonged discomfort (such as a mare with a retained placenta) given usually from a lateral recumbency position. A single sigh-like groan commonly occurs as a weary horse achieves recumbency. Some stabled individuals seem to use the groan while standing relaxed, as if bored with nothing to do. Most groans are audible only within a few meters of the source.

Blow

The non-pulsated, broad-band sound produced by forceful expulsion of air through the nostrils is called a blow (Figure 19.11i,j). Although some frequencies of these non-voiced sounds extend above 8 kHz, most of the sound energy is below 3 kHz. Blows are most audible within 30 m. When emitted as an expression of alarm (e.g., while hesitantly investigating a suspicious object several meters away), the average duration is less than 0.5 second (Table 19.1). Such brief and forceful blows apparently serve to alert nearby horses. Stevenson (1975) noted that nostrils dilated completely during the brief blow, the mouth remained closed, and lack of movement during and immediately after the sound was typical. More prolonged blows (range 0.6-1.3 seconds) are emitted during olfactory investigation when the individual exhales after a bout of sniffing.

Snort

Snorts are also broad-band sounds of forceful exhalation through the nostrils but are characterized by an audible flutter pulsation (Figure 19.11k). The nostrils can be seen to flutter with each pulsation and the mouth remains closed. The average duration of a snort is 0.8-0.9 second, and loudness is normally sufficient to hear the sound at 50 m. Horses emit snorts when the nasal passage is irritated (such as with dust), sometimes immediately after vigorous locomotion, or when the individual is restless and yet constrained, such as by a human handler or barrier. Under the latter conflict situations, snorts appear to be a displacement activity and seem to express the horse's restlessness. A snort-like exhalation with rapid flutter occasionally occurs with labored breathing.

Snore

Snores are broad-band, raspy inhalation sounds (Figure 19.11 l). These non-voiced sounds seem incidental to inhalation under especially two circumstances. One is prior to emitting an alarm blow, where the preceding inhalation occasionally is a brief audible snore lasting 0.3-0.5 second. If such a sound functions in communication, it probably serves as a preparatory or sensitizing cue for the subsequent alarm blow. The second situation is with labored breathing of a recumbent horse, when the inhalation may sound like a human snore lasting 1.0 to 1.8 seconds (Table 19.1).

Other Sounds

Hoofbeats may have a function in equine communication. The sounds can indicate the presence and location of individuals plus the type of locomotion being used. The accentuated hoofbeats of a horse circuitously investigating a suspicious object appear to gain the attention of other horses.

Incidental sounds of eating, tail switching, coughing, grooming, snapping, shaking, and so on possibly convey information to neighboring horses about on-going activity. Specialization of these sounds for communication is not apparent.

TACTILE INTERACTIONS

When two horses interact at close range, tactile exchanges often occur. Upon initial greeting during naso-nasal interaction, some direct touching may occur as well as the indirect tactile effects of forceful exhalation. One or both individuals may then make contact at the flank or genital region of the other. The importance of such tactile activity is not clear.

Mare-foal interactions often involve tactile activity. Mothers nudge their foal periodically with their muzzle to direct the foal's movements. Nuzzling with the upper lip also occurs, apparently to offer reassurance. A prolonged bout of licking by the mare occurs soon after parturition but rarely occurs thereafter. Foals nibble and lick their mothers, especially in the first day; nudging and sucking at the teats occur during nursing. The foal's nuzzling at the flank and ventral surface of the mare seems to signal care solicitation.

Foals sometimes induce their mother to interact with them in a bout of mutual grooming by first nibbling at the mare. Allogrooming in older horses is normally initiated by one individual gently nibbling at the neck or withers of another.

Aggression often involves tactile interactions. Biting, pushing, striking, and kicking are tactile signs of aggressiveness. The intensity of the signals reflects the seriousness of the interaction. In mock fighting, the approach, sequence of events, intervening activities, and level of intensity apparently cue opponents that the interaction is playful.

Tactile cues are also exchanged between horse and human handler. A rider can interpret much about a horse's coordination, tractability, attentiveness, and understanding of commands by utilizing tactile signals transferred, for example, via the reins and the rider's legs. Numerous tactile signals are likewise given to the horse by the rider; only a portion may be intentional commands. Various pieces of horse apparatus, such as bits and spurs, are often designed and used for tactile effect.

CHEMICAL EXCHANGES

Olfactory cues seem to be sought by horses as they approach and investigate each other during their rather ritualized greeting interaction. Sniffing is obvious at the initial naso-nasal and head phase; it often is continued at the flank and genital region. Excrement and novel objects are typically investigated extensively using smell. Flehmen sometimes occurs. How much information a horse gains from olfactory investigation is not known; however some discrimination between individuals apparently occurs, as evidenced by marking behavior of stallions and the ability of mares to distinguish their own foal using olfaction. When Wolski et al. (1980) modified olfactory cues between mare and foal, the individuals had difficulty finding their appropriate partner.

Stallions appear to be initially attracted to mares in estrus primarily by visual, rather than olfactory, cues. Once in contact with a mare, the stallion may proceed with olfactory investigation and with testing the receptivity of the mare. If the mare does not object, most stallions continue to show sexual interest, especially toward adult mares, without additional evidence of olfactory investigation. When the mare shows slight signs of objection, the stallion may smell the mare's vulva and urine. Flehmen may follow, and in some cases loss of libido occurs. Wierzbowski (1959) found adult stallions with experimentally impared olfaction showed no inhibition of sexual behavior. Inexperienced stallions, however, showed more need for odor cues; for example young stallions showed sexual interest in a dummy if it was first sprinkled with urine of an estrous mare. Experience, therefore, seems to play a role in how extensively and when a stallion uses odor cues.

Excrement, saliva, breath odor, secretions, and numerous glandular areas (Schaffer 1940) on the skin of horses provide possible sources of olfactory cues. Further study will be necessary to better understand chemical communication in horses.

Part VI

Applied Ethology

Behavioral Symptoms

Behavior can be a valuable tool when judging the health and well-being of a horse. Problems typically have a behavioral component whether those problems are physiological or psychological. Thus a behavioral symptom can be used as a signal that there has been an alteration in the horse that may need medical attention or other special care. A change in behavior is often determined by comparing observed patterns to traits shown previously by the individual or by most horses of the same age and sex.

Behavioral symptoms include atypical postures and facial expressions, decreased ability to orient or to move in a normal manner, apparent loss in perception, poor maintenance activities, altered social interactions, and excessive agonistic behavior (Figure 20.1). In some cases, only a shift in intensity, frequency of occurrence, or rhythmicity is symptomatic. At other times, entirely new behaviors occur. The appendix is a listing of many behavioral symptoms and the possible problems they indicate.

CHANGES IN EXPRESSION AND POSTURE

Expression and posture changes, indicative of a variety of maladies, can involve not only the head and neck but also the rest of the body. Thus an abnormal angle or atypical movement of the legs, tail, back, neck and head should be noted with concern. Peculiarities in the way the ears, eyes, eyelids, lips, tongue, lower jaw, and nostrils are positioned or moved may signal problems, such as localized irritation or neural damage involving branches of the cranial nerves.

Some expressions signal pain (e.g., Walser 1965, Fraser 1969). For example, discomfort in the abdomen is often indicated by pawing, rolling, staring at flanks, groaning, frequent lying down and getting up, bumping the belly by lifting a hindleg, or sitting dog-like with forelegs extended supporting the forequarters. Discomfort in the shoulder, pelvis, or limbs is manifested as lameness.

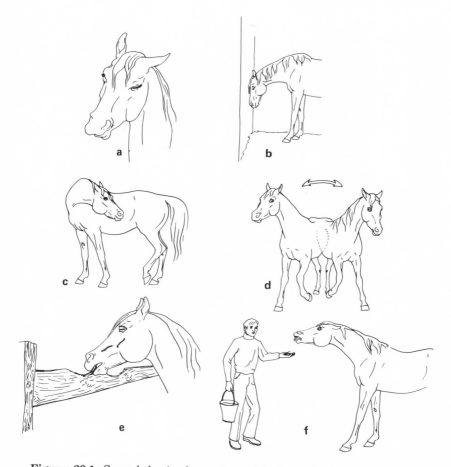

Figure 20.1: Some behavioral symptoms: (a) change in facial expression (damage to left facial nerve), (b) change in posture or orientation (head press against wall), (c) sign of discomfort (looking at abdomen), (d) stereotyped movement (weaving), (e) pica (wood chewing), (f) change in temperament or social behavior (abnormal aggression).

Some ailments exhibit characteristic postures or expressions while others vary in their effects on the behavior of individuals. For example, tetanus typically causes a rigid, spread-leg stance with head and tail extended; encephalomyelitis may be signaled by an abnormal stance, drowsiness with yawning, drooping lower lip, or oscillations of the eyes (Byrne 1972, Knight 1972, Siegmund 1973).

CHANGES IN PERCEPTION AND ORIENTATION

Some ailments cause changes in the perception or orientation of horses (see

Appendix). For example, certain toxins and diseases such as periodic ophthalmia or encephalomyelitis can impair vision and can cause disorientation. The visually impaired individual typically fails to investigate new objects and may collide with them; the horse may stumble, or exhibit a cautious high-stepping gait, or when led may move the ears excessively. Shying and other forms of fear may at times be due to impaired vision. Nerve damage can cause not only loss of vision but also hearing and olfactory impairment, loss of reflexes, and localized anesthesia. Circling, vertigo, and aimless wandering are behaviors that result from any of a variety of problems such as poisoning, brain lesions, and infections that impair the horse's orientation (Siegmund 1973).

CHANGES IN MOTOR COORDINATION

Motor coordination symptoms appear with many types of problems. Abnormalities of locomotion (such as staggering, stiffness, and lameness) can result from toxins, infections, and a variety of anatomical or physiological problems with the legs and feet (cf. Rooney 1981). Tremors and clonic muscle spasms (rhythmic contraction) can occur with ailments such as paspalum fungus poisoning, epilepsy, rabies, and the convulsive syndrome of newborn foals; tonic spasms (continuous tension) appear, for example, in tetanus, meningeal disease, mucormycosis, eclampsia, and with certain poisons such as lead or strychnine (e.g., Siegmund 1973).

Incoordination, lethargy and weakness appear with many ailments, including heat exhaustion, respiratory disease, cirrhosis of the liver, and severe hypoglycemia. In some cases, horses show a reluctance to move and may have paralysis. Damage may have occurred to peripheral nerves or involve dysfunction of central nervous tissue. Yet a horse suffering a ruptured stomach, laminitis, azoturia, or exhibiting an agonistic response to a handler may also refuse to move.

Lameness frequently is signaled by a change in the way the head and neck move during locomotion. Abnormal head bobbing appears. When a foreleg is affected, the head and neck drop when the healthy foot lands and raises as the painful foot makes contact. In hindleg lameness, the opposite pattern occurs—the head raises when the sound foot lands and drops when the lame leg makes contact.

CHANGES IN MAINTENANCE BEHAVIOR

Symptoms of many maladies appear as changes in resting pattern, ingestive behavior, respiration, grooming, eliminative behavior, and other maintenance activities. Dental problems usually alter characteristics of chewing, whereas ailments of the pharynx or esophagus can affect swallowing and the interest of the horse in food or water. Stressful environments can affect ingestion patterns also. Coughing and alterations of the respiratory cycle, including nostril dilation, can result, for example, from infections, toxins, heat exhaustion, respiratory or esophageal obstruction, heaves or other respiratory ailments. Foals may lose the sucking reflex with various septicemias and bacteremias as well as with the convulsive syndrome of newborn foals (Rossdale 1968b; Siegmund 1973).

Ailments can affect the way a horse conducts comfort behaviors and cares for itself. Skin ailments may cause excessive rubbing and result in open wounds. Ear parasites may cause frequent head shaking. Grooming may be prolonged, unusually frequent, or neglected entirely. Rolling can be frequent in a horse experiencing abdominal pain. Colic appears with many ailments, such as with allergic reactions, toxins, ruptured bladder, scrotal hernia, parasitic infections, and a variety of gastrointestinal problems. Calculi in the urinary system and cystitis affect urination patterns. Sweating occurs excessively with some ailments, such as gastritis, severe hypoglycemia, and allergic reactions; yet, sweating stops with salt deficiency and heat exhaustion.

[handwritten margin note: Signs of abdominal pain]

CHANGES IN SOCIAL BEHAVIOR

[handwritten note: usually show dumb form]

Some ailments noticeably alter social behavior of horses (see Appendix). For example, increased aggression characterizes rabies. Changes in sexual behavior can be caused by such factors as nutritional deficiencies and malfunction of the gonads. The tendency for horses to become solitary can appear in acute infections and toxicosis; solitary tendencies seem characteristic of locoweed poisoning and the convulsive syndrome of foals. It is not unusual, however, for mares about to give birth to leave their social group to achieve temporary isolation.

Horses that as young foals are isolated for a prolonged period from other horses may show little desire for equine companionship later, especially if an alternate form of companionship was available during the isolation period. If a human or another species serves as foster parent and companion during a foal's otherwise isolated early life, social preferences tend to focus on members of the foster parent species rather than on horses (cf. Grzimek 1949a). Subsequently, when pastured with other horses, the individual with altered social development prior to weaning will tend to exhibit solitary tendencies when the foster species is not present. The foal's sensitive period for primary socialization is evident during the second hour postpartum; where the period ends is not known (Waring 1970b). Prolonged social contact over many days accentuates in foals the effect of primary socialization, thus social preferences become entrenched.

APPEARANCE OF PROBLEM BEHAVIORS

Sometimes, in what may seem to be a healthy horse, behavioral patterns appear that interfere with the utility, well-being, or esthetic value of the individual. These behaviors are often called vices and include such behaviors as pica (chewing and ingestion of unnatural items), cribbing (Figure 20.2), weaving, bucking, bolting, shying, and rearing (e.g., Temple 1963). The behaviors are not without cause and can be symptoms that the horse is suffering, for ex-

∧, cause?. or result?

ample, from a dental problem, nutritional deficiency, parasitic infection, or is being housed or handled improperly. Unless the cause of the behavioral problem is corrected soon after the pattern appears, the behavioral trait may become a habit and persist. Temporary inhibition of the undesirable behavior can sometimes be achieved with special equipment. However, treating only the symptoms may eventually lead to other problems unless the fomenting cause of the abnormality is also corrected.

Figure 20.2: Cribbing by pushing upper incisors against a fence post. Often accompanied by swallowing air (aerophagia or windsucking).

When horses are removed from the open range as well as their normal social environment and when they no longer can spend much of their day foraging and expending energy seeking suitable resources, the likelihood is high that abnormal traits will appear. Feeding of concentrated foods further complicates the situation by eliminating the additional time confined animals can occupy themselves with manipulating and ingesting food.

Confined horses with restless energy may begin stall walking, digging, weaving, wood chewing, cribbing, or kicking of the stall walls to relieve boredom or frustration. Varied environmental experiences, companionship, roughage diet, and regular program of exercise help reduce the occurrence of these behaviors. Returning problem horses to a pasture situation should be considered (Houpt 1981).

Oftentimes sexual behavior problems can be reduced, it not eliminated, by proper nutrition or by careful regulation of sexual interactions. Anestrous mares frequently become cyclic with improved nutrition. A stallion used too frequently for breeding and one that experiences pain or other adverse events during coitus may develop impotence or become unable to complete intromission with ejaculation. In some cases, ejaculation is inhibited only in certain situations, for example by an artificial vagina but not during natural mating (Bielanski 1960). Behavioral inhibitions can often be reversed by correcting the cause and using conditioning techniques to re-establish libido, mounting, and successful intromission.

Disturbances during copulation which prevent ejaculation may induce excessive biting by the stallion. A stallion repeatedly affected appears to generalize and becomes unduly aggressive toward each mate. Tyler (1969) observed that stallions upon being turned out onto open range from winter confinement attempted to copulate with mares whether or not they exhibited estrus. Once a mare was selected, a stallion's pursuit was relentless. The stallions became increasingly aggressive and in some instances attacked young mares and fatally mauled foals. The aggressiveness of the stallions eventually waned as the breeding season progressed and sexual behavior became oriented primarily toward mares in estrus. Stallions remaining on the range over the winter did not show the unusual aggression or the indiscriminate interest in mares.

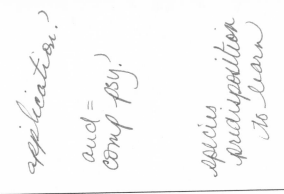

21

Learning and Memory

Learning is fundamental to the survival of horses. Although a horse has innate abilities, they alone will not suffice; it is the adaptive modification and the supplementation of those traits that make an individual successful.

The capacity to learn has been one of the features that made the horse ideal for domestication. Extensive training is possible. And once skills are learned, retention is prolonged. Yet horses vary. One of the needs of the horse industry is for a reliable measure of the trainability of horses before considerable time and expense are devoted to an individual. It is apparent that individual differences do occur; some horses have more potential than others.

Prior experiences greatly influence the behavior of horses. They are particularly affected by experiences that cause pain or fear. As trainers well know, even one bad experience, such as with new equipment or at a particular location, will result in a horse showing anxiety each time the same or similar situation reappears. In some cases, the memory of the experience seems to last for years. Considerable training is often required to overcome such negative experiences. Positive experiences, on the other hand, facilitate subsequent interactions and learning. For example, neonatal foals that have been extensively handled readily overcome fear responses to new stimuli and show far more independence of the mother as well as greater exploratory tendencies than unhandled foals (Waring 1972).

Although the process of learning may be a continuum, investigators have found it convenient to divide the phenomenon into several categories. The so-called "types" of learning in the classification scheme include habituation, classical conditioning, instrumental conditioning, imprinting, latent learning, insight, and imitation.

HABITUATION

The first type of learning evident in newborn foals is habituation. This trait, apparent soon after birth, involves the reduction of a response upon repeated stimulation. A neonate, for example, will soon cease to withdraw from tactile stimulation and will then allow body contact, such as gentle grooming by man or the mother, without objection. Repeatedly throughout its life a horse habituates to stimuli that are frequent and of no consequence. In this way, the individual adapts to initially frightening noises, objects, and many other stimuli that regularly appear in its environment. Sometimes stimulus generalization occurs where the adaptation is shown even to stimuli that are somewhat similar but not necessarily identical to those encountered before.

Initial training of a horse regardless of its age often involves habituation. The horse must adapt to close human contact, to the apparatus used during training, as well as to features of the training site before training can effectively proceed to other levels of learning. Many trainers begin working with naive horses by first exposing them to stimulation caused by and associated with the trainer. Tactile, auditory, and visual stimuli are repeatedly directed at the horse in a way that fear and aggressive responses become noticeably diminished through habituation.

CLASSICAL CONDITIONING

While a horse generally learns to ignore frequent stimuli that are of little consequence in themselves, the individual also learns that some initially inconsequential stimuli (CS) are regularly associated with stimuli (US) that trigger a response. Subsequently, the horse begins to give its response as soon as the CS appears without the prior dependence of the response on the US. This development of a new stimulus-response association is called classical conditioning. The CS is the conditioned stimulus, and the US is the unconditioned stimulus. In classical conditioning, the horse's response does not necessarily alter the occurrence or sequence of subsequent environmental events.

Examples of classical conditioning in horses are common but little studied. Foals that require periodic medical treatment soon learn a click of their stall door latch is associated with the human intruder, thus they commence to show a withdrawal response even before the door opens and the intruder appears. Horses in stables also learn prefeeding sounds and activity; those stimuli then release responses, such as begging, initially seen only after food itself appeared. During training and handling, horses often learn to anticipate commands and changes in activity because of the associated conditioned stimuli inadvertently given by the handler. Anticipation becomes evident also when environmental events occur at regular intervals, such as at a particular time of day. When the factor of time is paired repeatedly with events, such as the appearance of the caretaker, the horse associates the time with the event and begins to show anticipatory watchfulness.

INSTRUMENTAL CONDITIONING

In instrumental conditioning, the behavior of the horse influences the sequence or occurrence of subsequent events. Typically the horse's response leads to some degree of reward or punishment. Thus a horse learns to respond so as to bring about reward and not punishment. Whether by trial and error or by the manipulation of a handler, horses thus learn to open covered boxed to obtain food, to press a lever to activate a watering device, to respond to commands, to distinguish between similar items, and to do or not do various other activities.

The capacity of horses to apply instrumental learning to their daily life can be illustrated by a 23-year-old mare I studied at the University of Munich (Waring 1974). Although the mare often drank and ate in a typical manner, she and two other horses in the barn would occasionally begin a session of dunking hay in water before ingesting it. The mare's trait was to lift the hay from the floor pile by sliding large amounts up a nearby wall and onto a concrete shelf at the rear of her stall. The hay was pushed along the shelf with her muzzle until it was against a small water basin attached to the rear wall. The mare then took small amounts of hay in her incisor teeth, dunked the hay into the water basin, and chewed. More than one dunking often occurred before the hay was swallowed and the process repeated. With each dunking the pressure plate of the self-watering device was activated, thus soon water overflow had the shelf, wall, and floor saturated.

The observed hay-moistening behavior was not a stereotyped behavior or done without purpose. Wetting the hay was the motivation. Fresh cut grass was neither moved to the shelf nor dunked in water. Hay soaked for one hour in water and fed to the mare also did not induce the regular dunking trait. Yet as soon as dry hay was present, the mare commenced the routine of dunking the dry material in her water basin an average of 5.1 times per minute as she ate. When the self-watering device was turned off and no water was available in the basin, the dunking trait waned (extinguished) as shown in Figure 21.1. Recovery of the behavior pattern promptly occurred upon my restoring water flow to the device. To test if the trait was unique to the self-watering device I extinguished the response to the water basin, then two buckets containing water were placed in a depression of the shelf. The mare soon began to steadily dunk hay in the buckets (Figure 21.2) and seldom tried the water basin. However, as soon as the horse saw the valve manipulated to restore water flow to the self-watering apparatus, she shifted her hay moistening behavior to the water basin exclusively. Thus, the mare exhibited considerable ability to use knowledge she had acquired.

Stabled horses provide frequent opportunities to witness the operant capabilities of the species. In our horse research barn at Southern Illinois University, covers had to be installed over toggle as well as push-button switches within reach of the dexterous upper lip of certain horses who acquired the ability to activate the switches. Furthermore, double locks have been necessary on some stall doors to dissuade the departure of those individuals who have learned to grasp and lift the original latch with their teeth. Koegel (1954) reported about a gelding who periodically removed a bar from his stable door. Egress occurred to join a mare outside the stable.

Oftentimes horses have the ability to gain access to covered food containers by using their mouth, upper lip, or muzzle. In tests with horses, Gardner (1933) found acquisition of a technique to open a covered feed box was rapid and was

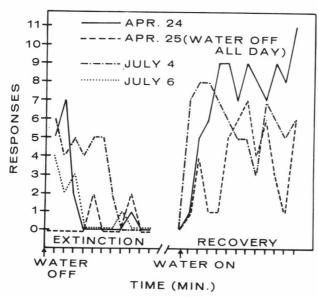

Figure 21.1: Experimental extinction and subsequent recovery of a conditioned response characterized by the horse dunking mouthfuls of hay in the basin of a self-watering device. The response waned when water was no longer present and recovered when the horse saw the water control valve opened. (Waring 1974).

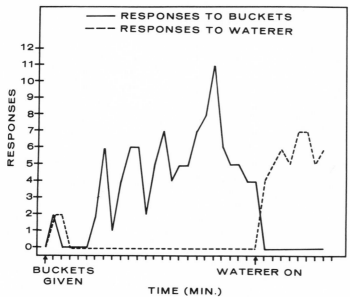

Figure 21.2: Development of hay-dunking responses to a new water source following extinction to a self-watering device, then the total shift to the original source when free choice was provided. (Waring 1974)

perfected in 3-4 trials, especially in the 5-14 year age group. In some cases, retention of the learning was still evident 6 to 12 months later.

Occasionally horses learn to wield objects at other horses apparently in play; a crude form of tool use. Dark (1972) watched a full-grown gelding repeatedly lift, aim, and toss a wooden pole in the direction of another horse (see Figure 5.1b). Gertrude Hendrix (pers. comm.). similarly observed a yearling gelding repeatedly lift a rubber feed pan and, while holding the pan in his teeth, he approached and spanked a yearling filly who was trying to graze. The filly eventually became aggressive and ended the companion's game which had recurred on two successive days.

Numerous experimental procedures have been developed to study instrumental learning. These include free-operant conditioning and procedures using discrete trials, such as avoidance learning and discrimination learning. Most experiments with learning in horses have used variations of these procedures. Reinforcement, if administered, is either after every correct response (continuous reinforcement=CRF), after several correct responses have been performed (fixed ratio=FR and variable ratio reinforcement=VR), or following the first correct response after an elapsed period of time (fixed interval=FI and variable interval reinforcement=VI).

Experimentation using a lever-pressing apparatus has been applied to horses. Similar to other animals tested for free-operant responses, Myers and Mesker (1960) found a horse gave relatively stable rates of response when several FR and FI reinforcement schedules where used. Evidence of anticipation appeared with FI schedules.

Hamilton (1911) studied the trial and error reactions of an 8-year-old gelding presented with four exit doors. In each trial, only one door could be opened; yet seldom would the horse try different doors to find the correct exit. In 86% of the responses, the horse focused its effort at one or two locked doors. Williams (1957) noted a similar tendency of horses to persist at one site when confronted with a detour problem (Figure 21.3). Alternate solutions are not readily attempted.

Most studies of learning in horses have used a discrimination problem. Gardner (1937a) confronted horses with three covered boxes; the horses learned the box that contained grain was always draped with a black cloth. When the cloth marker was then suspended low in front of the correct box, the average number of errors for 44 subjects doubled compared to trials 11 to 22 of the original paradigm. When the cloth was suspended above the feed box, discrimination errors quadrupled (Gardner 1937b). In another experiment when a 12-quart pail was used as a signal instead of a black cloth, the subjects (n=56) showed similar trends; errors were most frequent when the discrimination signal was suspended above the correct feed box (Gardner 1942).

Nobbe (1974) conditioned a horse to alternately nudge two rectangular polyhedrons (one black and the other white) suspended at nose height from the ceiling and spaced one meter apart. Grain was used for reinforcement. The response was taught (shaped) in two 15-minute sessions over a period of two days. After seven more sessions of the same length and a shift from continuous to fixed ratio (BWB or WBW) schedule of reinforcement, the horse was then required to give a BWBW response (FR4 reinforcement schedule) before being rewarded. In the first FR4 session 244 alternated responses were given; 292 occurred in the second FR4 session. Throughout the experiment, the response rate increased steadily even when an interval as much as one week occurred between sessions.

Figure 21.3: Horses tend to persist at one location rather than try alternate solutions to problems requiring a detour.

Warren and Warren (1962) required a pair of horses to learn to alternate between two feed boxes (black was on right and the white box was on left) after the horses had previously learned to seek hay from just one of the two boxes. Subsequent trials alternated between the two types of tasks (single versus alternate). Both subjects learned the successive reversal problem quickly. One horse averaged fewer than two errors per reversal over the series of nine tested; the other horse averaged two errors per problem during six reversals. There was a rapid decline in the number of errors made on consecutive reversals.

Voith (1975) expanded this type of research into a spatial reversal problem (where position was relevant, not the stimuli themselves) and a visual reversal problem (where stimuli were important, not their position). Black and white stimuli were used. The horses demonstrated progressive improvement in their ability to learn either type of reversal problem, although visual discrimination reversal problems seemed to be more difficult to learn than spatial.

Pattern discrimination learning has also been studied. Giebel (1958) conditioned a horse, a donkey, and a zebra to discriminate the correct choice in each of 20 pairs of patterns so as to obtain a food reward (Figure 21.4). The horse learned all 20 pairs, the donkey learned 13, and the zebra learned to discriminate 10 of the pairs. During a memory test at the end of training where each pair reappeared randomly a total of 30 times, the horse gave a perfect performance on four pairs and a performance low of 73% on one pattern. Retests at three, six, and twelve months showed virtually no memory loss on at least 19 of the pairs. Dixon (1966) conducted a nearly identical study with a 7-year-old pony gelding and found similar results. Retests at one, three, and six months showed an 11.5% loss of learning in the first month but little (3.5%) over the next five months. The frequency of correct choices on all 20 pairs at six months was 77%.

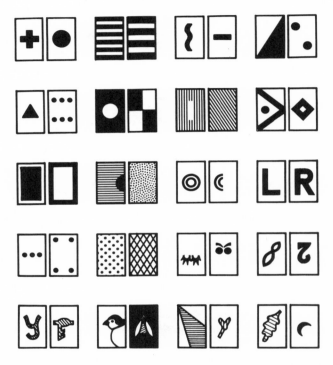

Figure 21.4: Pattern discrimination pairs used in the studies of Giebel (1958), Dixon (1966), and Voith (1975). The left pattern of each pair was the correct choice to obtain a food reward.

The learning set phenomenon described by Harlow (1949) seemed to be clearly demonstrated by the pony Dixon (1966) taught to discriminate patterns. The gelding learned how to learn. Successful discrimination of the first pair of patterns required numerous trials, but as the pony learned the game rules (i.e., that one stimulus of each pair led to a reward) fewer errors were made. From the sixth pair through the 20th the horse learned in one or two trials.

Voith (1975) repeated the pattern discrimination experiment, primarily to test for learning set and memory as well as to control for possible cueing by the experimenter. A trend of progressive improvement in learning successive pairs was demonstrated, although not as distinctly as in Dixon's study. In a non-reinforced retest immediately after all 20 pairs were learned, Voith's two mares achieved a 77% accuracy. When one mare was retested 12 months later, the performance was sluggish and was no better than chance. Reinforcement did little to improve the performance. Although the horse no longer demonstrated accurate pattern discrimination, it did remember how to work the experimental apparatus.

Mader and Price (1980) designed their discrimination study so horses had to learn to choose the correct visual stimulus (a checkerboard pattern) from a set of three stimuli. Among the 16 horses compared for their learning score, Quarter Horses learned faster than the Thoroughbreds tested. Learning performance declined with age. No relationship was found between dominance status and learning ability. Suggestive of the learning set phenomenon, learning progressed more rapidly for the second discrimination task than for the first.

Horses apply learned discriminations in their daily activity, for example, mares identify their own foal (e.g., Leblanc and Bouissou 1981). While grazing, many horses carefully choose and sort vegetation to obtain mouthfuls of specific plant species. Learning appears to be involved in such traits. Marinier (1980), while investigating selective grazing, found horses could be easily conditioned to avoid one of two equally palatable plant species. Two kinds of plants were repeatedly presented to each experimental horse and mild punishment was administered when a wrong choice was made. Although discrimination was learned, the horses differed in the number of trials needed to achieve success and in the amount of punishment required.

Popov (1956) reported his experimental horses could discriminate very slight changes in acoustical, visual, and tactile stimuli. Such signals are common in horse training. To test for the ability of horses to respond appropriately upon auditory, visual, and tactile cues, Yeates (1976) constructed a horse-size lever-pressing device. In the box chamber, each of three mares was taught to push a hinged flap to obtain food reward. Each mare learned the task within 1.5 to 2 hours. The horses then had to learn that food reinforcement would only occur when a flap-pushing response was done in the presence of either a yellow light, a coarse-sounding buzzer, or a pulsating tactile stimulus remotely applied at a forerib. By the end of a 21-day period, each mare had an improved performance, yet individual differences were evident. One mare was then left in the chamber continuously under a free-operant situation cued by the visual stimulus only. Her performance in five days had improved from 66.9% to 94.4% correct responses.

In another study designed to measure the learning ability of horses, Fiske and Potter (1979) applied the serial reversal discrimination technique of Warren and Warren (1962) on 26 yearling Quarter Horses. Mean trials (MT) and mean errors (ME) required to achieve criteria were computed for each horse, then a relative learning ability index (LAI) was calculated (1000/MT/ME). A single subjective trainability score (1 to 6) was obtained from a trainer. Linear regression analysis revealed a reduction in MT and ME ($P<0.01$) over the 21-day test period indicating learning set formation. Differences ($P<0.05$) were evident between sexes for MT and ME. Disruption of mental concentration due to estrus was suspected, at least for several fillies. Significant correlation between trainability score and learning ability measures (MT, ME, LAI) was evident for colts and geldings but not for fillies.

A two-component maze also has been used to assess the learning ability of horses. As a subject enters the first compartment of such a maze, it must turn either right or left around a partition to enter the second compartment. One route leads to an exit, the other does not. Kratzer et al. (1977) used the maze in a study of 37 yearling geldings. When a right-side choice was required during five trials, both latency of escape and errors decreased. Then a left-side choice was required of the horses. Latency and errors again decreased, but after three trials these values were still relatively high. The horses tended to still try the right side. Thus an adversive stimulus (CO_2 fire extinguisher discharge) was presented whenever a horse started to enter the dead-end compartment. Errors subsequently decreased during the remaining three trials, but latency did not. Test subjects fed 10%, 13%, 16%, and 19% protein diets did not show consistent differences in learning ability. In other studies using a two-compartment maze as well as a shock-avoidance technique, Haag and her coworkers (1980) found no correlation between dominance rank and learning ability.

To achieve learning, training can be frequent or spaced with long intertrial intervals. Rubin et al. (1980) conducted a study to look at the effect of such temporal distribution of training sessions. Horses were taught to respond in a particular manner upon the presentation of a visual or auditory cue so as to avoid receiving a mild electric shock. Some horses received daily training, a second group had training twice a week, a third group had one training session per week. The horses trained once a week achieved the learning criteria in significantly fewer sessions than subjects trained daily; yet the elapsed time from start of training to completion was greater because training was spread over many weeks. The twice-a-week group learned at a rate intermediate to the other two experimental groups.

Grzimek (1949c) investigated memory in horses using variations of a delayed reaction experiment where food was hidden in one of 3 or 4 boxes as a horse watched. After a slight delay, the horse was allowed to make its choice. The procedure eventually adopted was to have either of two subjects stand four meters from an array of four adjacent choice boxes. Grain was overtly dumped into one box, the handler then stepped behind a screen, and after the scheduled delay the horse was given the command "come." One horse, a mare, achieved the correct choice only up to six seconds of delay; at 15 seconds, her performance was at the level of chance. The other horse, a gelding, achieved a delay of as much as 60 seconds before his performance approached that expected by chance. A yearling filly tested by Nobbe (1978) using a two-choice delayed reaction procedure and a modified Wisconsin General Testing Apparatus achieved a delay of 24 seconds with accuracy above 80%. Unfortunately, the study had to be terminated before longer delays could be tested.

In summary, experimentation with instrumental conditioning of horses has shown horses can master numerous discrimination tasks as well as maze and avoidance learning. In some cases, memory can be prolonged. Horses, when provided several potential alternatives at once, tend to concentrate their trial-and-error efforts at only one or two of the alternatives. Habit strength develops rapidly. Individual differences in learning do occur; some breed differences may

not enough data to make such sweeping concl.

occur, but more data are needed to confirm those results. Differences between sexes are not consistent, as is the effect of emotionality on learning. Youthful horses, but not necessarily the youngest, perform better in learning tests than older horses. Dominance rank and learning ability are not correlated. Although teaching a horse an entirely new task may be tedious, similar tasks are learned with considerable improvement. Horses learn in fewer trials when the sessions are spaced at intervals rather than concentrated into a short time span.

IMPRINTING

Imprinting is a special type of learning where a long-term association or attachment is acquired during a limited stage in an animal's life. The sensitive period for social imprinting is often early in the life of the organism. In foals, this sensitive period is first evident during the second hour of age (Waring 1970b). A foal typically forms its initial social attachment to its own mother and hence its own species; however, Grzimek (1949a) reported that a foal isolated from its own species for the first 64 days exhibited a social preference for only its human companions when given free choice. It has not been determined if such a horse would maintain its cross-species preference into adult life nor what behaviors would be affected in maturity.

Not only does imprinting seem to occur in foals, a similar phenomenon appears in mares soon after parturition. Each mother rapidly learns to distinguish her own foal during a sensitive period that begins at parturition and apparently lasts no more than a few days. Once the mare has developed an attachment for a foal during the sensitive period, it is difficult to get her to accept any other neonate.

Other types of imprinting, such as food or habitat imprinting, are possible but have yet to be demonstrated in horses.

LATENT LEARNING, INSIGHT, AND IMITATION

Latent learning is the association of stimuli or situations without obvious reward or the trait itself being evident at the time of learning. It often results when an animal becomes familiar with its surroundings. For example, features about the geography and many objects in the environment can be learned during exploration; at the time, the individual shows no evidence that it has received positive feedback or reward. But later, the individual may apply what it has learned in a way that enhances its survival. Thus a horse new to its range may return undeviatingly at mid-day to the shade of a lone tree it had passed earlier while exploring.

While working with horses, handlers periodically witness that a horse had developed awareness and abilities beyond those being taught or that are exhibited only later after training has ceased. Williams (1957), for example, noted a novice mare she was training to jump showed little progress prior to a severe illness and a 9-month rest. Yet the individual returned to training with awareness and ability noticeably beyond that a learning curve would have predicted had training continued 9 months earlier. The mare had apparently acquired associations with earlier training that had yet to be fully assimilated before the prolonged rest.

Hendrix (1968) witnessed what may have been latent learning and concept formation. One day she had decided to reschool a proper canter to flighty show-ring horse by using four level stretches of terrain each with a steep hill at the end. The steep slope would hopefully allow her to regain control if the horse broke into a racing gallop. The first two level stretches were negotiated success-fully with a walk, trot, walk inserted between the places to canter. As the third flat stretch came into view the horse began to snort, collected its head, and moved in an excited manner. The anticipatory horse showed awareness that at level stretches the command to canter could occur. Another example Hendrix experienced was with a horse that had repeatedly been required to pause at a roadside. The command to proceed across the road was given only when oncom-ing traffic subsided. One day when the rider thought the way was clear she urged the horse forward, but the horse refused. The horse took heed of an on-coming car and did not proceed until it had passed. The horse seemed to have associated stopping at the edge of the road with oncoming traffic, not simply a whim of the rider.

Insightful problem solving, where an individual uses a combination of two or more learned tasks to solve a new problem, has not been systematically in-vestigated in horses. Again, anecdotal evidence could be cited, such as the rela-tively intricate schemes some horses devise to rid themselves of a novice rider. The hay-moistening mare mentioned earlier in this chapter may have been using insight to adapt to the use of buckets when the watering device was turned off. The ability for insight is likely; yet in most daily activities and in most problems they encounter, horses show little that can be attributed to in-sightful problem solving.

Learning through imitation is sometimes attributed to horses that begin cribbing or weaving. However, experimental evidence is lacking that vices are acquired by imitating others. Glendinning (1977) reported orphan foals did not graze until turned out with older horses. Marinier (1980) found trial-and-error was more likely responsible for selective grazing in foals than was imitation. Undoubtedly some behavioral traits that horses acquire are learned more quickly because of the following along with experienced herd members. Moving toward and using a new water hole is an example. Yet as with other forms of learning, considerable research is needed before a definitive explanation about the acquisition of the numerous learned traits of horses can be given.

22

Behavioral Manipulation and Restraint

Numerous techniques have been devised for manipulating the behavior of horses. Rarely are experimental data available as to their effectiveness. The intent here is not to review or critique the many techniques but to emphasize basic concepts of effective handling as supported through empirical evidence and to provide some guidance not only to achieve success but also to minimize stress for both the horse and handler.

Stress can often be kept to a minimum by gradually exposing a horse to new stimuli and situations. For example, the frequency of occurrence, intensity, speed of onset, and duration of new stimuli or novel situations can usually be controlled and innocuously established over several training sessions rather than initially at full strength. Habituation and learning set formation can be applied. With repeated stimulation and experiences, adaptation proceeds and the horse tends to adapt to subsequent new situations with increasing ease. Csapó (1972) applied these steps when weaning foals. By gradually and sequentially altering the close association of mare and foal, weaning was accomplished without the physiological and psychological stress shown by the abruptly weaned foals of the control group.

By reducing the number of factors causing stress, manipulation is made easier. Thus training is typically conducted in a familiar, rather than strange, environment. The stress induced by transporting horses can be reduced by providing conditions with favorable noise levels, temperature, air flow, humidity, and physical space as well as secure footing and stablized air pressure within aircraft. Orientation parallel to the line of travel seems to cause less restlessness upon starting and stopping than a transverse position. Cregier (1979, 1981) concluded that facing horses away from the direction of travel in trailers and vans greatly reduces physiological as well as psychological stress, especially during braking. This necessitates some design modifications in forward-facing equipment. To achieve balanced transport, a horse in the rear-facing position must be permitted room to raise, lower, and turn its head. Backing the horse into a rear-facing trailer compartment circumvents the fear horses have of walking into a darkened area.

The ease and effectiveness of behavioral manipulation and restraint can be greatly enhanced when a handler takes into account the innate tendencies, prior experiences, and learning of the horse. For example, a swift and focused approach toward an unhandled horse can initiate a flight response at a considerable distance. The withdrawal effort and flight distance will be less if the approach is slower and more indifferent. As repeated approaches occur and the horse experiences no adverse impact from the approaches, the result will be a diminished flight response. Further steps to habituate the horse to human interaction can then be instituted without applying restraint. A horse that associates an approaching handler with unpleasant experiences will subsequently show avoidance. In such a case, counter conditioning with a program of reinforcement can be an effective procedure to overcome the avoidance trait. Although time and patience are required, the risk to horse and handler is minimal.

Reflexes can inadvertently hinder or be used to advantage during behavioral manipulation (cf. Rooney 1981). Postural reflexes can induce either extension or flexion of the legs. To more easily persuade a naive foal to pick up its left hindfoot, an assistant can activate vestibular and neck reflexes by turning the head to the right and also lifting the head and neck slightly with a hand under the chin. The right legs extend but the left tend to flex thereby aiding the lifting of the left hindfoot. Similarly, lifting of the left forefoot can be induced by a right head turn, but this time the neck is flexed ventrally by hand pressure on the nose (Rooney 1979).

EARLY EXPERIENCE AND HUMAN SOCIALIZATION

Beginning human contact early in the life of a foal can potentially establish a close relationship with man that may facilitate subsequent handling and training (e.g., Marwick 1967). The sensitive period to establish initial social relationships in a foal commences very soon after birth (Waring 1970b). It is well known that in puppies primary socialization can be accomplished to both dogs as well as human handlers (e.g., Pfaffenberger and Scott 1959, Fox 1965). Both passive human exposure as well as active handling procedures were found to be effective in dogs (Stanley and Elliot 1962, Stanley 1965). The potential exists that foals, too, can establish a long-term association to their own species as well as to humans during their sensitive period. To pursue this idea, I exposed newborn American Saddlebred foals to various degrees of human contact beginning as early as 5 minutes to as late as 15 hours postpartum. To avoid imprinting foals exclusively to man, an effort was made to not disrupt the development of the mare-foal relationship.

The results were encouraging but not conclusive. Our 1969 experiment can serve as an illustration. Two foals were separated from their mother 5 minutes after birth and actively human fondled until returned to their mother at 70 minutes of age. A third foal received passive human exposure by a person sitting in the foaling stall from 1 to 6 hours postpartum. A fourth foal was bottle fed at hour 5 and 6 and had a human mannequin (Figure 22.1) in the foaling stall until 83 hours of age. The fifth and sixth foals were exposed to the manne-quin for 40 and 84 hours, respectively.

Figure 22.1: Mannequin used to provide newborn foals with a type of passive human socialization. (Waring 1970b)

To test the effect of their early experience, five foals (I-V) were individually halter-led away from the barn for 10 to 15 minutes during their first day. The handler directed and reinforced the foals using tactile and vocal stimuli. A sec-ond handler led the mother nearby. A similar halter leading session was conducted the second day postpartum. Following the second session, the foals were neither led nor extensively human handled until three months of age. At three months of age, all six foals were haltered and led outdoors.

The comparative responses of the foals during the test sessions are shown in Table 22.1. During the first halter session, difficulty was encountered primarily with foal V, the individual that lacked actual human interaction. During the second session foals I, III, and V were difficult to halter; yet leading was success-ful with each of the five foals.

During the session at three months of age, all foals objected to being haltered. Only foal I was tractable; foals II-V remained obstinate and objected to handling throughout the session. Foal VI, which was handled for the first time, acted fearful and confused yet did not show the same tenacity shown by foals II-V.

Table 22.1: Treatment and Comparative Responses of Six Foals to Various Types of Early Human Socialization

Type of Early Experience:

(a) Active Handling — Foals I and II were separated from dams and human fondled from 5 to 70 minutes postpartum

(b) Passive — Foal III received passive human exposure during hour 1 to 6; foal IV was bottle fed at hour 5 and 6

(c) Mannequin — Foal IV with mannequin from birth until hour 83; foal V, until hour 40; and foal VI, until hour 84

Responses Scored During Haltering and Leading	. . .Day 1. . .					. . .Day 2. . .					. . . .Day 90. . . .					
	I	II	III	IV	V	I	II	III	IV	V	I	II	III	IV	V	VI
(a) Avoided stationary human	o	o	o	o	o	+	o	+	o	o						
(b) Objected to being haltered	o	o	o	+	+	+	o	+	o	+	+	+	+	++	+	++
(c) Objected to walking with handler	o	o	o	o	+	o	o	o	o	o	o	+	+	++	+	+
(d) Tended to back or bolt while on lead line	+	o	o	o	+	o	o	+	o	o	+	+	+	+	++	++
(e) Tended to rear and fall	o	o	o	o	o	o	o	o	o	o	o	++	++	+	++	o
(f) Objected to handler's directions	o	o	o	o	+	o	o	o	o	o	o	+	+	++	+	+
(g) Uneasy at end of session	o	o	o	o	+	o	o	o	o	o	+	++	++	++	++	++

*Responses scored as follows: o = no clear response, + = response overt, and ++ = response strongly overt.

Data from Waring 1972

It is difficult to separate the effects of early human interaction from such factors as heritable tendencies, maternal influences, or prior experiences resulting from management procedures inadvertently unique to each mare and foal. Nevertheless, human socialization both active and passive seems to have an effect on newborn foals. We were interested to observe after the first halter session that foal II, after release in a paddock with the mare, soon returned to the vicinity of the handler and slept in lateral recumbency within two meters of him. At the second session, foal III after release in the paddock followed behind the handler until herded away by the mare.

A foal's working relationship with the handler at the first halter session seemed dependent upon the degree of previous human exposure. The more handled foals performed better. By the second session, initial avoidance of the handler was appearing in foals without some form of continued exposure to a human form; yet once haltered all the foals were manageable. When human socialization was discontinued for three months, the advantage of early human interaction diminished. The quality as well as quantity of socialization during a sensitive period are undoubtedly both important to establish long-term associations.

The foals we have human handled during the early hours after parturition have not had their relationship with the mare greatly disrupted. However, compared to unhandled foals, the handled foals quickly habituate fear responses, are bolder, and exhibit more exploratory behavior. They leave their mother's side more readily and to greater distances when first turned outdoors, approach other organisms, and show more self-confidence. These behavioral differences characterize these foals as they continue to develop. Such behavior causes the mare to spend more time following her foal and herding it away from contacts with others. Unhandled foals are more reluctant to leave the side of their mother. The early handled foals are, therefore, subject to more dangers resulting from their zealous curiosity and diminished inhibition. *brkdwn of maternal bond?*

One advantage of establishing the early foal-human relationship is the ease the newborn foal shows toward handling. Such a foal is little stressed. Furthermore, communication readily develops and training can progress effectively. Heird et al. (1981) found that handled horses scored higher on learning tests than individuals not handled in the first year of development.

During early handling it is physically easy to teach a neonatal foal to relax upon restraint. To establish such training, the foal is repeatedly restrained (Figure 22.2) and subsequently released only upon its relaxation. Such learned relaxation can be valuable when later in life the horse becomes entangled accidentally in wire, equipment, or other hazard. It will tend to await assistance without struggling. *data?*

Figure 22.2: Restraining technique useful on a newborn foal.

While investigating early experience, we found as human-socialized foals develop they tend to treat human handlers more and more as conspecifics and peers. Conspecific-oriented play and aggression of foals can be dangerous to a handler. Although similar conspecific-like activity exists in a dog-human relationship (Fox 1965), little danger occurs to the handler because of the relative gentleness of the behaviors and small size of the pup. As a foal develops, it is not readily subjugated by a handler merely because of prior human socialization. The handler's dominance must be re-exerted, at least with juveniles, at nearly every session.

Although more research is needed on early experience and human socialization, it appears early handling provides a means to develop a foal with certain traits, such as an ability to more readily control fear responses, habituate promptly to new situations, exhibit self-confidence, and interact with its surroundings. The desirability of developing a horse with such characteristics must be evaluated by each horse owner to best coincide with the future utility planned for the individual.

FUNDAMENTALS OF TRAINING

An effective trainer knows not only the behavior of horses in general but also the specific traits of the horse being trained. Inherited traits as well as those acquired through prior experience will enhance or hinder training. Sometimes traits frustrating to one trainer will be used to advantage by another. Some horses learn faster than others and to more advanced levels. Athletic abilities also vary as do emotionality, attentiveness, and sensitivity to signals from the handler. Training should follow a systematic plan of building upon some traits and modifying others. Along the way, adjustments in the plan will be necessary to compensate for the individual's abilities and progress.

A suitable environment for training is important. Progress normally occurs most effectively in situations where fear responses and distracting stimuli are minimal, and where correct responses can be best ensured. Habituation to the environmental situation can be an important first step in a training program. The horse is allowed to adapt to the surroundings through repeated or continuous exposure to the fear-inducing or distracting stimuli. For example, a horse may require a period of getting accustomed to a handler or to being attached to tack and other equipment before the individual is attentive to further training. A habituation procedure used by some horsemen is sacking out, where the inexperienced horse is rubbed and contacted repeatedly with a cloth or saddle blanket before each attempt to saddle the horse.

When seeking a new response, it is prudent to structure the situation to facilitate the correct response rather than some undesired response. An arena, for example, reduces the risk that a horse will get out of control. Aids such as a martingale, a barrier, a corner, a curved turn, or an assistant can be used effectively to facilitate correct responses. Since horses tend to repeat previous responses, a special effort to have the response correct the first time is worth a few extra precautions.

Training often builds upon existing stimulus-response relationships. For example, innate tendencies to move away from some stimuli and avoid others can be used effectively to develop new stimulus-response associations. Thus, while free schooling or longeing a horse, a handler can initially use the position of his body to induce such responses as forward locomotion or stopping. Based on the innate flight response, closing in on the horse from behind will cause the horse to move forward. If the handler then shifts so the horse moves progressively closer to the handler, the horse is induced to stop. A whip can be used to extend the handler's body in a desired direction without need for much change in actual location. Commands to move forward or to stop can be paired during training with the already effective stimuli and the horse will learn to respond to the appropriate visual or acoustical signal. The commands should immediately precede the already effective stimuli during the paired conditioning trials.

Commands should be specific for each response desired and given in a manner the horse can clearly discriminate from other signals. Walking, stopping, and turning are commonly taught before advancing to faster gaits and more complex maneuvers. Many trainers prefer to longe and ground drive their horses before advancing to training under saddle.

Progressing gradually and sequentially from basic familiar activities to the more difficult facilitates understanding between horse and handler. The horse becomes attuned to signals emanating from the handler, and the handler can more easily learn the capabilities of the horse. The phenomenon of learning set formation (see Chapter 21) can be used to advance otherwise difficult training.

Successful horse training is often contingent upon reinforcement and its timely application. In many cases, reinforcers are used that the horse has learned, such as a kind voice. Reinforcers that tended to reward the horse are called positive reinforcement, whereas stimuli aversive or noxious to the horse are considered negative reinforcement. To be effective in establishing or squelching a particular stimulus-response relationship, the reinforcement should immediately follow the response. This is especially important with negative reinforcement since the horse will associate the reinforcement with whatever it has just done. If some correct response occurs just as a negative reinforcer is administered, the horse will be reluctant to give the correct response again. Reward following misbehavior will tend to establish that behavior. Thus, undesirable behavior is often brought on by the untimely and unjudicious use of reinforcement, positive as well as negative.

Horse training typically involves a mixture of positive and negative reinforcement. A reassuring voice and gentle stroking with a hand are examples of positive reinforcement; release from aversive situations can be rewarding as well. Negative reinforcement includes the harsh use of a bit, spurs, quirt, whip, or a chiding voice. Pleasant as well as unpleasant events occur during handling sessions; yet handlers tend to emphasize one system of reinforcement over another. More often negative stimuli are emphasized using such techniques as avoidance conditioning, escape conditioning, or punishment (Fiske 1979). With the latter, the horse is reprimanded for making a response in the absence of a cue or for failing to recognize a cue. In escape conditioning, the horse receives a noxious stimulus and must learn how to respond to eliminate the stimulus. A harshly applied bit can be such a case. In avoidance conditioning, the horse learns that if it responds appropriately to a signal it will avoid receiving an aversive stimulus that would otherwise occur. Using exclusively a positive reinforcement system with a small dose of dilute maple syrup remotely supplied to the mouth of a 2-year-old naive gelding, Carl Pitts (pers. comm.) quickly established verbal commands, riding, jumping, and assertive behavior (where the horse was trained to be more aggressive to other horses in a pasture).

The technique of shaping can be used to guide a horse into a proper response. With this method the individual is reinforced for successive approximations until the end response is achieved. A general response is first accepted as an approximation and is rewarded, the next reinforcement is given immediately after a better approximation, and so on until the horse does the sought-after response.

Continuous reinforcement is usually used during initial training, but once learning has occurred an intermittent reinforcement schedule is commonly applied, at least for positive reinforcement. Performance is less likely to extinguish under an intermittent reinforcement schedule than if the organism were suddenly to get no reinforcement after having gotten rewarded previously on every occasion.

Attentive and responsive horses learn most efficiently, thus trainers usually try to maintain motivation and avoid boredom as well as overwork. Lessons commonly range from 5 to 15 minutes, rarely more than 30. Several brief lessons will normally achieve more than a single long session. Habits form and anticipations occur when sessions are routine.

Ending a training session should be viewed by the handler as a form of positive reinforcement. Thus, it is important to end following a set of correct responses by the horse. To avoid rewarding undesirable behavior when a horse has misbehaved or has not been able to achieve a new task, it may be necessary to return to previously learned activity before stopping a handling session. For this reason, some horse owners find it worthwhile to reschool their horse at least briefly following handling by another person.

When undesirable behavior appears in the routine of a horse, treatment early can usually alter the undesirable trait more easily than waiting until the behavior is well established through habit. In some cases, simply not reinforcing a response will cause it to diminish through the process called extinction. Unfortunately most undesirable behaviors are not so easily eliminated. For example, vices brought on by boredom seem to be self reinforcing. The underlying cause of misbehavior should be determined and corrected, as warranted. Lack of companionship, fear, excess energy due to insufficient exercise, improper handling, illness, trauma, or nutritional deficiencies may be involved. Correct diagnosis of the situation will forestall making matters worse. A new environmental setting is often helpful. Punishment is not an effective way to treat fear-induced responses, such as balking at a stream crossing or kicking when a leg is handled. Harsh treatment may temporarily inhibit the trait but will likely compound the real problem (Voith 1979a). Treatment must involve eliminating the fear.

Modifying an undesirable behavior can follow the same principles used in training. Usually the horse must learn that alternate responses are more appropriate, and the trainer works toward achieving specific alternatives through conditioning techniques using reinforcement (positive especially). Occasionally habituation is required to desensitize an individual to fearful situations. Repeating fear-inducing stimuli frequently and especially with a gradual increase to full intensity will cause the individual to adapt to the stimuli as it learns that no harm follows. Patience and persistance are required of the handler in any training procedure; these virtues are indispensible when correcting behavior problems since hundreds of trials, if not thousands, may be required. Short cuts may not be available or may instead be considered too risky to pursue.

RESTRAINT

Casting or the severe technique of choking down a horse has occasionally been used on wild, intractable horses to permit flooding the individual with human handling. For example, Catlin (1857) observed American Indians successfully subjugated feral horses in a single session by physically casting a horse to a recumbent position and then progressively handling the individual (especially around the head) before allowing it to stand. Succinylcholine chloride has been used in a similar way to cast unruly, aggressive horses and permit a barrage of handling while the horse is immobilized but fully conscious of stimulation (Miller 1966). The overt objection the unruly horse has toward human contact before restraint and handling is substantially diminished after such extreme treatment. This type of treatment causes the horse to become submissive to the handler. Follow-up handling is used to maintain the willingness of the horse to permit human contact and manipulation.

Physical and chemical restraint are usually used to reduce risk (1) to humans, (2) to the horse itself, or (3) to other horses while some momentary procedure is carried out. In most instances, training is neither intended nor deemed a practicable alternative to assure safety and reduce the likelihood of unwanted movement. Restraint is an attempt to counteract the horse's innate tendency to take flight or to exhibit aggressive responses during contact. Numerous techniques for restraint have been devised (e.g., Leahy and Barrow 1953, Fraser 1967, Catcott and Smithcors 1972).

Oftentimes little restraint is applied to a horse that has learned through prior experience to accept handling of various kinds. Yet even a well-mannered horse may suddenly show a vigorous agonistic response when frightened or in pain. Precautionary measures can be valuable. Nevertheless, applying a method of restraint more severe than is necessary for a particular procedure is also undesirable; the horse may object more to the restraint than to the procedure itself. Since the apparatus and manipulations used in restraint can themselves cause uneasiness in a horse, it is useful to conduct practice sessions where the stimuli are gradually and repeatedly presented to habituate fear responses.

Physical restraint once applied should be held firmly and with confidence. Horses under weak and inadequate control tend to become increasingly restless and unruly. Occasionally a horse learns to resist one type of restraint and a new method must be employed. Hand holding of restraining apparatus is commonly practiced to permit rapid release in an emergency. Quick-release fittings, operable even under tension, should be used whenever fixed lines are used. The surroundings, including the substrate, should be chosen and prepared to facilitate safety, the pending restraint, and the required procedures.

At the minimum, the head is normally restrained if no more than with a hand or line to a halter (head collar). Some handlers additionally apply a chain shank of a halter line either over the muzzle, under the jaw, through the mouth, or over the upper gums (gingiva) to gain greater control (Vaughan 1972). Bits and other pieces of tack can function to aid restraint.

The war bridle (or rope gag) is a simple and effective method of restraint when used prudently. A rope is passed over the poll and either over the upper gingiva (Figure 22.3) or around the lower jaw (Figure 22.4). Pull on the rope causes localized discomfort which commonly results in a general immobility response.

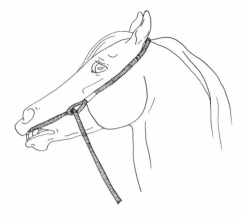

Figure 22.3: Variation of a war bridle where rope pressure is applied to the gingiva above the upper incisor teeth.

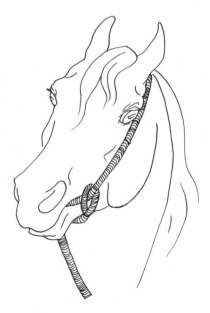

Figure 22.4: War bridle where constricting pressure occurs to the jaw as rope tension is increased.

The twitch (Figure 22.5) is commonly applied to cause head restraint and body immobility. Many commercial and homemade twitches are used with varying degrees of effectiveness. The principle of each is to apply pressure to the sensory nerves of the lip, typically the upper lip near the incisor teeth. The discomfort produced diverts the horse's attention while treatment is conducted elsewhere on the horse's body (Leahy and Barrow 1953). Grasping an ear or a handful of skin at or behind the shoulder can, in some cases, cause similar immobility.

Figure 22.5: The twitch is one of the more commonly used methods of restraint.

Occasionally the head must be restrained from turning to lick a wound on the body or legs. A side stick (Figure 22.6) will prevent the horse from reaching its hindquarters but does allow grazing and foreleg contact. A cradle (Figure 22.7) will prevent a horse from reaching both the fore- and hindquarters.

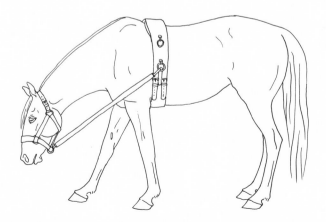

Figure 22.6: A rigid side stick can be used to prevent a horse from licking or biting at wounds on the hindquarters.

Figure 22.7: A cradle limits turning as well as lowering of the neck. It is used to prevent a horse from gaining access to wounds on its body or legs. (Adapted from Leahy and Barrow 1953)

Various kinds of hobbles are used to restrict leg movement. For example, a strap or rope may be used to bind a single foreleg bent at the knee (Figure 22.8). Or both forelegs may be fastened together with a short strand of webbing or leather cuffed around each pastern. Breeding hobbles limiting backward movement of the hindlegs are used to restrain a mare from kicking during breeding or gynecological procedures. Such hobbles can be fashioned from rope, leather, or webbing; some, such as the one in Figure 22.9, permit walking yet restrict kicking.

Figure 22.8: Stirrup-strap hobble used to restrict locomotion while treating a horse. (Adapted from Leahy and Barrow 1953)

Figure 22.9: Breeding hobble used on mares to prevent kicking; nevertheless, with this design, walking is possible.

With reasonably manageable horses, manual restraint of a fore- or hindleg can be performed by lifting and holding the pastern or cannon. For an injection into a lower hindleg, manual restraint can be achieved by lifting the leg from the opposite side and directing it beneath the belly and forward of the supporting hindleg.

The tail can also be used to aid restraint. Manual displacement of the tail over the horse's back or to one side is often effective to discourage kicking and permit rectal or urogenital examination. A rope tied to the tail just at the end of the last coccygeal vertebra permits a solid tail attachment that can be used for hindquarter restraint or even to help support a hindleg for treatment. In the latter instance, the rope is passed through a strap circling the hind pastern then passed over the horse's back to an assistant standing on the opposite side of the horse (Figure 22.10).

Cross-ties, fixed lines attached to each side of the halter, are commonly used to restrict the movement of a horse during grooming, saddling, and standard management procedures.

Stocks are devices usually made of wood or metal that are designed to restrict lateral as well as forward and backward movement of a standing horse while permitting access for medical treatment. For example, two rows of posts may be used, or a set of crossbars supported by corner posts. The length and width of the space within a stock is designed to just fit the length and width of the horse.

Slings are normally used to help support a horse in a standing position. To avoid physiological complications the horse should be able to support its own weight at least partially. Occasionally a sling is used to temporarily assist a horse to its feet. A broad belly band with an additional band at the chest, another around the hindquarters, and all linked to a single support from above can form a satisfactory sling (Vaughan 1972).

Skittish horses can oftentimes be quieted by covering their eyes with blinders to restrict or prevent visual stimuli.

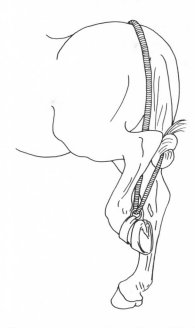

Figure 22.10: Tail-rope technique to hold up a hindleg for treatment.

Occasionally it is necessary to force a horse into a recumbent position. A pony or small horse can be cast manually by standing on one side of the animal, reaching over the back for the halter and tail (passed forward by an assistant through the groin to the opposite flank), then by turning the head backward and pulling upward with both hands the animal is brought down against the handlers legs. Casting harnesses are usually used for large horses. Various designs exist and are commonly made of rope, leather, or webbing. To accomplish casting, the hindlegs are drawn under the horse by pulling on side lines. In the double side-line casting harness (Figure 22.11), one person pulls the rope on one side, another simultaneously pulls the other side-line in the opposite direction, and a third person holds the halter rope to guide the fall and to restrain the head once the horse is prone. Subsequently the legs can be bound in a flexed or in an extended position.

Chemical agents are now commonly used to cast horses. Caution is taken to use a halter and line to prevent the horse from falling backward into a wall or against some stationary object. In some cases, a drug that causes muscle relaxation but without loss of consciousness (e.g., succinylcholine) is used. At other times one or a combination of anesthetics may be administered.

Under the influence of a general anesthetic, loss of the ability to stand occurs in the stage of anesthesia called Stage II. This sometimes follows a period of unrest or even excitement, depending on the temperament of the horse, the environmental situation, as well as the drug used. Stimulation must be kept minimal even in Stage II to avoid struggling and excitement. Reflexes in this stage become exaggerated, nystagmus often occurs (sometimes accompanied by blinking), pupils dilate, the ears may twitch, muscle tone is increased, and breathing is irregular. As anesthesia advances, the signs of Stage III (Surgical Anesthesia) are seen beginning at Plane 1. The horse no longer responds to painful stimuli. Pupils become constricted, except with a few drugs. The palpebral, anal, and corneal reflexes are present in light surgical anesthesia (Plane 1) but become weak and absent as anesthesia becomes deeper. Breathing and heart beat become steady, and muscle relaxation occurs. In the deepest plane (Plane 4) of surgical anesthesia, signaled by pupil dilation and dribbling of urine, both respiration and cardiovascular function are severely depressed; thus risk of death is high if the animal is not soon returned to a lighter level of anesthesia. This deep level of anesthesia is therefore avoided.

The duration and nature of recovery from anesthesia depends on the drugs used. Precautions should be taken during recovery to protect the horse from injury due to uncoordinated movements. Struggling is more frequent in animals that have not received preanesthetic tranquilizers and in animals excessively restrained or stimulated during recovery (Gabel and Jones 1972).

Food and water should be withheld from horses recovering from general anesthesia, or when under other forms of chemical restraint, until full recovery occurs.

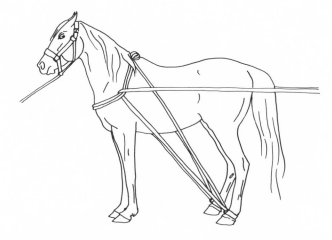

Figure 22.11: Double side-line casting harness used to force a horse to recumbency by pulling the side lines simultaneously in opposite directions. (Adapted from Leahy and Barrows 1953)

Tranquilizers are used to control nervousness and hyperexcitability in horses. The treated individuals become drowsy and less responsive to their surroundings; nevertheless, when subjected to strong or unusual stimulation, especially pain, they may react more violently than they would without a tranquilizer. Besides calming a horse, tranquilizers are sometimes used to induce prolapsing of the penis to permit cleaning and examination (Gabel 1972).

Sedatives are also used to quiet horses. These compounds cause a sleep-like state, but not necessarily recumbency, and usually reduce the horse's responsiveness to painful stimuli. Sedatives commonly show their effect more rapidly than tranquilizers. Sedatives or tranquilizers are commonly used as preanesthetic medication. Narcotic analgesic compounds tend to stimulate locomotor activity in horses (Combie et al. 1979) and are seldom used when restraint or quiescence is desired.

The effectiveness of chemical compounds for restraint is affected by such factors as dosage, manner of administration, and the physiological and psychological state of the animal. Emergency medical procedures are sometimes required to counteract unforeseen complications while using drugs. Without adequate knowledge and safeguards, use of chemical agents is not advised.

Appendix

Variety of Equine Behavioral Symptoms and Possible Problems Indicated

Behavioral Symptom	Dysfunction and Trauma	Nutrition, Allergy and Toxins	Parasites and Infections	Handling and Management
Expressions and Postures				
Head:				
Ear immobility	Facial nerve damage	—	—	—
Drooping ears	Hearing impairment	—	African horsesickness	Drug (depressant)
Rigid ears	—	—	—	Drug (stimulant)
Rapid ear movement while walking	Vision impairment	—	—	—
Oscillating eyes/nystagmus	Vestibular system lesion	—	Encephalomyelitis	—
Diverging eyes/strabismus	Oculomotor nerve lesion	—	—	Drug reaction
Staring/stupor	Discomfort, Cerebral edema	Senecio poisoning	—	—
Looking at flanks	Abdominal discomfort	—	—	—
Nictitating membrane closure	—	—	Tetanus	—
Unilateral eyelid closure	—	—	Periodic ophthalmia (uveitus)	—
Facial paralysis	Facial nerve damage	—	Infection of gutteral pouches	—
Grimace expression	—	Poisoning (yellow-star thistle, Russian knapweed)	—	—
Deviation of nose	Facial nerve damage	—	—	—
Drooping lower lip	Basal ganglion disease	—	Encephalomyelitis	—
Hypertonia of upper lip, hypotonia of lower lip	—	—	—	—
Curling of upper lip	—	Allergic reaction	—	—
Tongue and lip tremors	Basal ganglion disease Stomatitus,	—	—	—
Lolling of tongue	Paralysis of tongue	—	—	—
Involuntary chewing, tongue flicking	—	Yellow-star thistle poisoning	—	—
Champing of jaws	Convulsive syndrome of newborn foals	—	—	—

(continued)

Behavioral Symptom	Dysfunction and Trauma	Nutrition, Allergy and Toxins	Parasites and Infections	Handling and Management
Grinding of teeth	Esophageal disorder	Poison hemlock poisoning	—	—
Difficulties in chewing	Trigeminal nerve injury	—	—	—
Mouth remains open	Trigeminal nerve injury	—	—	—
Drops food	Facial nerve dysfunction	Yellow-star thistle poisoning	—	—
Immerses muzzle but does not drink	Eclampsia	Blister beetle poisoning (cantharidin)	—	—
Yawning, drowsiness	—	Yellow-star thistle poisoning, Senecio poisoning	Encephalomyelitis	
Head tilt	Vestibular system lesion			
Head pressing	Cirrhosis of liver, Serum hepatitis, Cerebral cortex damage	Poisoning (e.g., lead, Senecio, locoweed, moldy corn)	Gutteral pouch infection	—
Head shaking	Trigeminal neuralgia	—	Ear irritation (e.g., flying insects, spinose ear tick)	—
Jerky up and down movement of head	Convulsive syndrome of newborn foals	—	—	
Neck:				
Neck arching with retching (choking)	Obstruction of esophagus, Spasm of esophagus, Gastritis	—	—	
Neck lowering, head extension	Esophagitis	—	African horsesickness, Strangles	
Torsion of the neck	Vestibular system lesion	Dystrophic myodegeneration	Strangles, Tuberculosis	
Rigid neck	Intracranial meningitis	—		—
Abnormal nodding during walk	Unilateral/unequal foreleg or hindleg lameness	—	—	
Legs:				
Knocking/stamping	Distress	—	Chorioptic mange, Flying insects	—
Pawing	Arthritis of shoulder joint, Coxitis, Gonitis, Sweeney	—		—
Swing leg lameness		—	—	—

(continued)

Behavioral Symptom	Dysfunction and Trauma	Nutrition, Allergy and Toxins	Parasites and Infections	Handling and Management
Excessive angle at hock joint	Gastrocnemius muscle or Achilles' tendon rupture	—	—	—
Kicks at abdomen	Abdominal pain	—	Flying insects	—
Uneven croup raising during walking	Unilateral/unequal hindleg lameness	—	—	—
Body (miscellaneous): Abnormal curve of back	Ankylosing lesions of vertebral column	—	—	—
Tail stiff and extended	—	—	Tetanus	—
Nonretracted penis	Pubic nerve damage	—	—	—
Shivering	—	Chlorinated hydrocarbon poisoning	—	—
Rolling	Abdominal discomfort	—	Rabies	—
Rearing	—	Allergic reaction	—	—
Shelter seeking	—	—	Flying insects	—
Uneasiness/restless	Discomfort (e.g., impaction)	Blister beetle poisoning, Allergic reaction	—	Heat exhaustion, Transfusion reaction
Groaning	Discomfort	—	—	—
Frequent lying down and getting up	Distress	Blister beetle poisoning	—	—
Nervousness	Adrenal medulla tumor	Poisoning (e.g., bracken fern, locoweed)	—	Subnormal early experience, Poor handling
Hyperexcitability	—	Poisoning (e.g., salt, penta-chlorophenol, sodium fluoroacetate, organic phosphates)	Rabies	Drug reaction
Incessant walking	Serum hepatitis Peritonitis, Purpura hemorrhagica	—	—	—
Reluctance to lie down	—	—	—	—
Abnormal resting stance	Leg/foot discomfort, lameness	Fluorosis, Laminitis, Bracken fern poisoning	Encephalomyelitis	—

(continued)

Behavioral Symptom	Dysfunction and Trauma	Nutrition, Allergy and Toxins	Parasites and Infections	Handling and Management
Extreme lethargy	Liver malfunction	—	—	—
Rigid stance	Myotonia, Peritonitis, Meningitis	—	—	—
Rigid stance with head extension, saw horse stance	—	—	Tetanus	—
Stands with both forelegs forward	—	Laminitis in forelegs	—	—
Stands with both hindlegs forward	—	Laminitis in hindlegs	—	—
Stands with one leg forward (pointing)	Navicular disease	—	—	—
Dog-sitting posture	Abdominal pain	—	—	—
Perception Changes				
Lack of ear movement	Deafness	—	—	—
No reaction to sounds	Acoustic nerve dysfunction, Trigeminal nerve lesion	—	—	—
Unusually sensitive to loud sounds	Facial nerve paralysis, Vision impairment	—	—	Procaine HCl overdose
Sudden noise causes tonic spasms		Strychnine poisoning	Tetanus	—
Impaired vision	Electric shock (lightning), Serum hepatitis, Pituitary tumor, Brain stem lesion, Cerebral cortex lesion, Optic nerve lesion, Superior colliculi lesion, Ruptured bladder	Poisoning (e.g., lead, selenium, salt, Hypericum, moldy corn), Snake bite, Stachybotryotoxicosis, Riboflavin deficiency	Encephalomyelitis, Periodic ophthalmia	—
Extreme sensitivity to light	—	Riboflavin deficiency	Periodic ophthalmia, Equine viral arteritis	—
Extreme cutaneous sensitivity	Residual effect to lightning shock	Strychnine poisoning	—	—

(continued)

Behavioral Symptom	Dysfunction and Trauma	Nutrition, Allergy and Toxins	Parasites and Infections	Handling and Management
Loss of corneal reflex	Ophthalmic nerve paralysis	—	—	—
Facial and oral anesthesia	Trigeminal nerve lesion	—	—	—
Unable to locate food by smell	Olfactory nerve lesion	—	—	—
Orientation Change				
Aimless walking	Serum hepatitis, Convulsive syndrome of newborn foals	Poisoning (e.g., locoweed, moldy corn, Senecio, Crotalaria)	Neonatal actinobacillosis	—
Circling	Cirrhosis of liver, Cerebral cortex lesion (occipital lobe), Brain stem lesion, Vestibular system lesion, Superior colliculi lesion	Poisoning (e.g., lead, moldy corn)	Encephalomyelitis Coccidioidomycosis	Procaine HCl overdose
Loss of equilibrium/vertigo	Electric shock (lightning)	—	Foal septicemia	—
Coordination Change				
Reluctance to feed off the ground	—	—	Tetanus	—
Lameness	Pain or anatomical abnormalities of limbs, feet, shoulder or pelvis	Laminitis, Nutritional hyperparathyroidism, Ergotism, Fluorosis, Chronic selenium poisoning Sweet clover poisoning	Lymphangitis, Verminous thrombosis, Melioidosis, Coccidioidomycosis	Improper foot care, Leg damage during handling or exercise
Lameness primarily after exercise	Splints, Metacarpal or metatarsal fracture	—	—	—
Lameness primarily after rest	Bone spavin	—	—	—
Irregular gait	Hypertrophic pulmonary osteoarthropathy	—	Encephalomyelitis	—

(continued)

Behavioral Symptom	Dysfunction and Trauma	Nutrition, Allergy and Toxins	Parasites and Infections	Handling and Management
Short choppy gait	Osslets	—	—	—
High-stepping gait	Vision impairment	Lupine poisoning	—	—
Upward jerking of hindlegs with first steps	Stringhalt, Upward fixation of patella	—	—	—
Stabs feet into ground during trotting	Bone spavin	—	—	—
Shortened stride	Bucked shins	—	—	—
Hindleg suddenly pulled posteriorly just before contact in walking	Fibrotic and ossifying myopathy	—	—	—
Continuous foreleg running motions while recumbent	—	—	Encephalomyelitis	—
Toe dragging	Bone spavin	Senecio poisoning	Cerebrospinal nematodosis	—
Drags hoof	Patellar luxation	Poisoning (e.g., bromide intoxication)	—	—
Stumbling	Vision impairment, Pain in heels	—	—	Heat exhaustion
Falling	Vestibular system lesion, Epilepsy	Milkweed poisoning	—	—
Staggering	Equine incoordination	Allergic reaction, Ryegrass staggers, Poisoning (e.g., locoweed, death camas, moldy corn)	Anthrax, Cerebrospinal nematodosis, African horsesickness Babesiosis	—
Hindquarter trembling induced by backing	Shivering syndrome	—	—	—
Stiffness	Purpura hemorrhagica, Azoturia/tying-up	White muscle disease, Ryegrass staggers, Vitamin D deficiency, Poisoning (e.g., lead, selenium, sweet clover, fluoride	Toxicoinfectious botulism, Equine influenza, Tetanus	—

(continued)

Behavioral Symptom	Dysfunction and Trauma	Nutrition, Allergy and Toxins	Parasites and Infections	Handling and Management
Resists locomotion	Myotonia, Azoturia, Stomach rupture, Purpura hemorrhagica, Pleuritis,	Laminitis	Encephalomyelitis, Tetanus	Agonistic response to handling
Paralysis	Brain stem lesion Spinal cord injury, Medulla oblongata lesion	Poisoning (e.g., botulism, poison hemlock)	—	—
Posterior paralysis	Obturator nerve damage, Spinal cord lesion (lumbar)	—	Surra	—
Forelimb paralysis	Radial nerve damage, Spinal cord lesion (lower cervical)	—	—	—
Prostration/shock	Severe trauma, Cardiac failure, Internal obstruction	Anaphylaxis, Intoxication, Snake bite, Lead poisoning	Hypoderma larvae rupture	—
Dystonia	Basal ganglion disease			—
Opisthotonus	Meningeal disease	—	—	—
No tail movement	Cauda equina lesion	—	—	—
Tremors/muscle spasms	Stomach rupture, Serum hepatitis, Severe hypoglycemia, Epilepsy (grand mal)	Poisoning (e.g., moldy corn, milkweed, paspalum fungus)	Rabies, Toxicoinfectious botulism	—
Tonic spasms/convulsions	Eclampsia	Poisoning (e.g., lead, githagin, toxicoalgae, strychnine)	Tetanus, Mucormycosis	—
Clonic spasms	Convulsive syndrome of newborn foals	—	—	—
Jerky movements	Cerebellar disease (dysmetria)	—	—	—

(continued)

Behavioral Symptom	Dysfunction and Trauma	Nutrition, Allergy and Toxins	Parasites and Infections	Handling and Management
Incoordination	Basilar skull fracture, Eclampsia, Cirrhosis of liver, Severe hypoglycemia, Myelitis	Vitamin deficiency (A, thiamine), Poisoning (e.g., salt, bracken fern, Senecio, paspalum fungus, castor bean, moldy corn, milkweed)	Toxoplasmosis, Encephalomyelitis, Ehrlichiosis, Foal septicemia	—
Lethargy/weakness	Hemolytic disease of newborn foals, Neonatal diarrhea of foals, Cerebral cortex damage	Nutritional deficiencies, Snake bite, Poisoning (e.g., lead, selenium, sodium fluoroacetate, inorganic arsenic)	Respiratory disease, Anthrax, Toxicoinfectious botulism, Tularemia, Neonatal septicemias and bacteremias (e.g., naval ill), Coccidioidomycosis, Large strongyle infection	Heat exhaustion
Weight shifted to forelegs in standing	—	Lathyrism		—
Maintenance Behavior Change				
Loss of appetite	Hemolytic disease of newborn foals, Pharyngeal paralysis	Nutritional deficiencies (e.g., protein, minerals, vitamins), Urticaria, Poisoning (e.g., bracken fern, Senecio, locoweed)	Lymphangitis, Respiratory disease, Anthrax, Malignant edema, Leptospirosis, Strangles, Blastomycosis	—
Slow, deliberate ingestion	Pharyngitis, Lampas			
Mastication poor	Dental abnormalities	Fluorosis		—
Irregular chewing	Dental problem			—
Head tilted while chewing	Dental problem			—
Quidding (drops bolus of food)	Dental problem			—
Refuses to eat hard grain	Dental problem			—

(continued)

Behavioral Symptom	Dysfunction and Trauma	Nutrition, Allergy and Toxins	Parasites and Infections	Handling and Management
Difficulty in swallowing	Paralysis or physical interference in mouth or throat	White muscle disease, Poisoning (e.g., moldy corn, thallium, yellow-star thistle)	Encephalomyelitis, Listeriosis, Toxicoinfectious botulism	—
Stands with mouth open	Stomatitis			
Excessive salivation	Foreign body in mouth, Pharyngeal paralysis, Swallowing difficulties	Allergic reaction, Poisoning (e.g., cyanide, pentachlorophenol, toxic algae, organophosphates)	Rabies	Motion sickness
Refusal to drink	Pharyngitis		Strangles	—
Loss of suck reflex	Convulsive syndrome of newborn foals	—	Neonatal septicemias and bacteremias	—
Excessive drinking	Diabetes insipidus, Pituitary tumor	—	Vesicular stomatitis	Boredom/restless energy
Thirst	Enteritis, Pituitary tumor (pars intermedia)	—	Babesiosis	—
Coughing	Heaves, Obstruction of esophagus, Hypertonic pulmonary osteoarthropathy	Chronic bronchitis, Heaves, Crofton weed poisoning	Respiratory disease, Adenoviral infection, Foal pneumonia, African horsesickness, Bordetellosis, Bronchitis, Blastomycosis, Tularemia, Corynebacterium equi infection, Coccidioidomycosis	—
Labored breathing, respiratory distress	Heaves, Pharyngitis, Nostril paralysis, Convulsive syndrome of newborn foals	Allergic reaction (feedstuffs, bedding, pollen), Snake bite, Poisoning (e.g., lead, poison hemlock, cyanide, nitrite, dinitro compounds, organophosphates)	Equine influenza, Anthrax, Coccidioidomycosis, Mucormycosis	Heat exhaustion, Respiratory obstruction during treatment

(continued)

Behavioral Symptom	Dysfunction and Trauma	Nutrition, Allergy and Toxins	Parasites and Infections	Handling and Management
Rapid breathing	Severe hypoglycemia, Hemolytic disease of newborn foals, Eclampsia	Allergic reaction, Cantharidin poisoning	Lymphangitis, Corynebacterium equi infection	Heat exhaustion, Drug reaction (stimulant)
Inspiration hurried, nostrils dilated	Heaves			
Audible inhalation	Facial nerve damage, Laryngeal hemiplegia	—	—	—
Sweating	Gastritis, Rupture of large intestine, Eclampsia, Azoturia/tying-up	Laminitis, Cantharidin poisoning	—	—
Profuse sweating	Extreme discomfort, Severe hypoglycemia	Allergic reaction, Castor bean poisoning	African horsesickness	—
Sweating stops	Anhidrosis	Salt deficiency	—	Heat cramps, Heat exhaustion
Frequent urination or attempts	Cystitis, Urethral calculi	Allergic reaction, Poisoning (e.g., tannic acid, cantharidin)	Toxicoinfectious botulism	—
Slow painful discharge of urine	Vesical calculi			
Colic	Constipation, Scrotal hernia, Ruptured bladder, Gastritis, Volvulus, Impaction of large intestine, Enterolith, Peritonitis, Visceral tumor, Tension of spermatic cord	Allergic reaction, Poisoning (e.g., castor bean, chlorinated hydrocarbon, fluoroacetate, corn-cockle)	Coccidioidomycosis, Strongyles, Anthrax	—

(continued)

Behavioral Symptom	Dysfunction and Trauma	Nutrition, Allergy and Toxins	Parasites and Infections	Handling and Management
Weakness	—	Locoweed poisoning	Equine infectious anemia, Botulism	—
Rubbing	—	Allergic dermatitis from insect bites	Lice, Mites	—
Social Abnormalities				
Solitary tendencies	Pending parturition	Locoweed poisoning	—	Abnormal early social development
Loss of foal-mother bond	Convulsive syndrome foals	—	—	—
Seeks only human social contact			—	Social imprinting on humans
Docility	Pituitary tumor (pars intermedia)	—	—	—
Change in personality or temperament (e.g., hyperexcitability, lethargy)	Liver malfunction	—	—	—
Irritability	Hypoglycemia, Cryptorchidism	Vitamin D deficiency	Listeriosis, Rabies	Drug overdose
Aggressiveness	—	—	Rabies, Lice	Boredom, Poor handling, Sexual frustration
Viciousness	Pathologic nymphomania	—	Rabies	Sexual isolation
Increased libido	Cryptorchidism	—	—	—
Prolonged estrus	Irritation of clitoris by foreign body, Cystic ovary, Neuroendocrine disorder	—	—	—
Prolonged anestrus	Infantilism, Ovarian malfunction	Undernourishment, Obesity	—	Poor nutrition
Impotence	Pseudocyesis	Undernourishment	—	Adverse experiences during breeding, Frequent ejaculations

(continued)

Behavioral Symptom *Problem Behaviors (Vices)*	Dysfunction and Trauma	Nutrition, Allergy and Toxins	Parasites and Infections	Handling and Management
Refusal to work	Liver dysfunction	—	—	Heat exhaustion
Difficulty to bridle	Dental abnormality	—	Spinose ear tick	—
Stall walking	—	—	—	Nervousness/excess energy
Weaving	—	—	—	Procaine HCl overdose Nervousness/excess energy
Cribbing/windsucking	Dental abnormality	—	—	Boredom, Poor social development
Pica (e.g., wood, dirt)	—	Nutritional deficiencies	Parasitic infection	Boredom
Coprophagia	—	Need for roughage	—	—
Fence chewing	Dental abnormality	Need for roughage	—	Boredom/restless energy
Digging	—	—	—	Nervousness/excess energy
Masturbation	—	—	—	Excess sexual energy
Charging/savaging	Hormone imbalance (e.g., cystic ovaries)	—	—	Poor handling, Anxiety
Striking	—	—	—	Agonistic response to handling
Kicking	—	—	—	Agonistic response to handling
Rearing	—	—	—	Agonistic response to handling
Bucking	—	—	—	Agonistic response to handling
Bolting	—	—	—	Agonistic response to handling
Jibbing	Vision impairment	—	—	Agonistic response to handling
Shying	Vision impairment	—	—	Fear response, Agonistic response to handling

(Spector 1956, Blood and Henderson 1963, Catcott and Smithcors 1972, Siegmund 1973, Marinier 1980.)

Bibliography

Adams, O.R. 1966. *Lameness in horses.* 2nd edition. Lea and Febiger, Philadelphia.

Allen, W.R. and P.D. Rossdale. 1973. Preliminary studies upon the use of prostaglandins for inducing and synchronizing oestrus in Thoroughbred mares. *Equine Vet. J.* 5:137-140.

Altmann, M. 1951. The study of behavior in a horse-mule group. *Sociometry* 14:351-366.

Andersen, S.R. and O. Munk, eds. 1971. An extract of Detmar Wilhelm Soemmerring's thesis: A comment on the horizontal sections of eyes in man and animals. *Acta Ophthalmologica, Suppl.* 110.

Archer, M. 1978. Studies on producing and maintaining balanced pastures for studs. *Equine Vet. J.* 10:54-59.

Arthur, G.H. 1970. The induction of oestrus in mares by uterine infusion of saline. *Vet. Rec.* 86:584-586.

Arthur, G.H. 1975. Influence of intrauterine saline infusion upon the oestrous cycle of the mare. *J. Reprod. Fert., Suppl.* 23:231-234.

Asa, C.S. 1980. Sociosexual behavior in the domestic pony. Pages 59-70 in R.H. Denniston, ed. *Symposium on the ecology and behavior of wild and feral equids, 6-8 September 1979.* University of Wyoming, Laramie.

Asa, C.S., D.A. Goldfoot and O.J. Ginther. 1979. Sociosexual behavior and the ovulatory cycle of ponies *(Equus caballus)* observed in harem groups. *Hormones and Behavior* 13:49-65.

Asa, C.S., D.S. Goldfoot, M.C. Garcia and O.J. Ginther. 1980. Sexual behavior in ovariectomized and seasonally anovulatory pony mares *(Equus caballus). Hormones and Behavior* 14:46-54.

Azzie, M.A.J. 1975. Some clinical observations on the effect of an implant of oestradiol benzoate in brood mares. *J. Reprod. Fert., Suppl.* 23:303-306.

Bannikov, A.G. 1961. Special natural conditions of the biotope of the Przewalski wild horse and some biological features of this species. [In Russian] *Equus* I:13-21.

Baskin, L.M. 1976. *The behavior of hoofed animals.* [In Russian] Science, Moscow. 296p.

Belonje, P.C. and C.H. Van Niekerk. 1975. A review of the influence of nutrition upon the oestrous cycle and early pregnancy in the mare. *J. Reprod. Fert., Suppl.* 23:167-169.

Benirschke, K. and N. Malouf. 1967. Chromosome studies of Equidae. *Equus* I:253-284.

Berger, J. 1975. Behavioral ecology of feral horses *(Equus caballus)* in the Grand Canyon. Thesis, California State University, Northridge. 82p.

265

Berger, J. 1977. Organizational systems and dominance in feral horses in the Grand Canyon. *Behav. Ecol. Sociobiol.* 2:131-146.

Bielanski, W. 1960. *Reproduction in horses. I. Stallions.* Publ. No. 116. Institute of Zootechniques and Agricultural College, Krakow, Poland.

Bielanski, W. and S. Wierzbowski. 1962. 'Depletion test' in stallions. *Proc. 4th Int. Congr. Anim. Reprod.,* 279-282.

Blakeslee, J.K. 1974. Mother-young relationships and related behavior among free-ranging Appaloosa horses. Thesis, Idaho State University, Pocatello. 113p.

Blood, D.C. and J.A. Henderson. 1963. *Veterinary medicine.* 2nd edition. Williams and Wilkins, Baltimore, 1224p.

Boy, V. and P. Duncan. 1979. Time-budgets of Camargue horses: I. Developmental changes in the time-budgets of foals. *Behaviour* 71:187-202.

Boyd, L.E. 1980. The natality, foal survivorship and mare-foal behavior of feral horses in Wyoming's Red Desert. Thesis, University of Wyoming, Laramie. 137p.

Breazile, J.E., B.C. Swafford and D.R. Biles. 1966. Motor cortex of the horse. *Amer. J. Vet. Res.* 27:1605-1609.

Brentjes, B. 1969. Equidenbastardierung im alten Orient. *Säugetierk. Mitt.* 17:141-151.

Brentjes, B. 1972. Das Pferd im alten Orient. *Säugetierk. Mitt.* 20:325-353.

Burkhardt, J. 1947. Transition from anoestrus in the mare and effects of artificial lighting. *J. Agric. Sci.* 37:64-68.

Byrne, R.J. 1972. Neurotropic viral diseases. Pages 46-57 in E.J. Catcott and J.F. Smithcors, eds. *Equine medicine and surgery.* 2nd edition. American Veterinary Publications, Wheaton, Illinois.

Catcott, E.J. and J.F. Smithcors. 1972. *Equine medicine and surgery.* 2nd edition. American Veterinary Publications, Wheaton, Illinois. 960p.

Catlin, G. 1857. *Letters and notes on the manners, customs, and condition of the North American Indians.* W.P. Hazard, Philadelphia. 2 v.

Christopher, M. 1970. *ESP, seers and psychics.* T.Y. Crowell, New York. 268p.

Clutton-Brock, T.H., P.J. Greenwood and R.P. Powell. 1976. Ranks and relationships in Highland ponies and Highland cows. *Z. Tierpsychol.* 41:202-216.

Collery, L. 1969. The sexual and social behaviour of the Connemara pony. *Brit. Vet. J.* 125:151-152.

Collery, L. 1978. Social interaction in an equine herd. *Proc. 1st World Congr. Ethol. Appl. to Zootechnics, Madrid, Spain.* Vol. I(E-I-8):87-91.

Combie, J., J. Dougherty, E. Nugent and T. Tobin. 1979. The pharmacology of narcotic analgesics in the horse. IV. Dose and time response relationships for behavioral responses to morphine, meperidine, pentazocine, anileridine, methadone and hydromorphone. *J. Equine Med. Surg.* 3:377-385.

Cox, J.E. 1970. Some observations of an orphan foal. *Brit. Vet. J.* 126:658-659.

Cregier, S.E. 1979. Alleviating surface transit stress on horses. Paper presented to Canadian Federal Veterinary Inspectors, University of Saskatchewan, Saskatoon.

Cregier, S.E. 1981. Alleviating road transit stress on horses. *Anim. Regul. Stud.* 3:223-227.

Crowe, C.W., R.E. Gardner, J.M. Humburg, R.F. Nachreiner and R.C. Purohit. 1977. Plasma testosterone and behavioral characteristics in geldings with intact epididymides. *J. Equine Med. Surg.* 1:387-390.

Csapó, G. 1972. Szopós csikók fokozatos elvalácztáza. *Állattenyésztés* 21:279-288.

Dallaire, A. and Y. Ruckebusch. 1974. Sleep and wakefulness in the housed pony under different dietary conditions. *Can. J. Comp. Med.* 38:65-71.

Dark, G.S. 1972. Two horses at liberty and their social interaction concerning a pole. Unpublished manuscript. Southern Illinois University, Carbondale.

Dark, G.S. 1975. Expressions of horses. Thesis, Southern Illinois University, Carbondale. 48p.

Denniston, R.H. 1980. The varying role of the male in feral horses. Pages 93-98 in R.H. Denniston, ed. *Symposium on the ecology and behavior of wild and feral equids, 6-8 September 1979*. University of Wyoming, Laramie.

Dixon, J.C. 1966. Pattern discrimination, learning set, and memory in a pony. Paper presented at the Midwestern Psychological Association Convention, Chicago.

Dixon, J.C. 1967. Acoustic behavior of horses (*Equus caballus* L.): Intraspecific salience and physical characteristics of three classes of recorded vocalizations. Thesis, Case Western Reserve University, Cleveland, Ohio. 47p.

Dobroruka, L.J. 1961. Eine Verhaltensstudie des Przewalski-Urwildpferdes (*Equus przewalskii* Poliakov 1881) in dem Zoologischen Garten Prag. *Equus* I:89-104.

Dolan, J.M. 1962. Remarks on the Przewalski horses in the Zoological Garden of Prague. *Säugetierk. Mitt.* 10:136.

Douglas, R.H. and O.J. Ginther. 1972. Effect of prostaglandin $F_{2\alpha}$ on length of diestrus in mares. *Prostaglandins* 2:265-268.

Drummond, A.J., P.C. Trexler, G.B. Edwards, C. Hillidge, and J.E. Cox. 1973. A technique for the production of gnotobiotic foals. *Vet. Rec.* 92:555-557.

Duke-Elder, S. 1958. *System of ophthalmology. Vol. I. The eye in evolution*. H. Kimpton, London.

Duncan, P. 1980. Time-budgets of Camargue horses. II. Time-budgets of adult horses and weaned sub-adults. *Behaviour* 72:26-49.

Duncan, P. and P. Cowtan. 1980. An unusual choice of habitat helps Camargue horses to avoid blood-sucking horse-flies. *Biol. Behav.* 5:55-60.

Duncan, P. and N. Vigne. 1979. The effect of group size in horses on the rate of attacks by blood-sucking flies. *Anim. Behav.* 27:623-625.

Ebhardt, H. 1954. Verhaltensweisen von Islandpferden in einem norddeutschen Freigelände. *Säugetierk. Mitt.* 2:145-154.

Ebhardt, H. 1957. Drei unterschiedliche Verhaltensweisen von Islandpferden in norddeutschen Freigehegen. *Säugetierk. Mitt.* 5:113-117.

Ebhardt, H. 1962. Ponies und Pferde in Röntgenbild nebst einigen stammesgeschichtlichen Bemerkungen dazu. *Säugetierk. Mitt.* 10:145-168.

Eldridge, F. and Y. Suzuki. 1976. A mare mule--dam or foster mother. *J. Heredity* 67:353-360.

Epstein, H. 1971. *The origin of the domestic animals of Africa*. Africana, New York. 2 v.

Estes, R.D. 1972. The role of the vomeronasal organ in mammalian reproduction. *Mammalia* 36:315-341.

Fagen, R.M. and T.K. George. 1977. Play behavior and exercise in young ponies (*Equus caballus* L.). *Behav. Ecol. Sociobiol.* 2:267-269.

Feist, J.D. 1971. Behavior of feral horses in the Pryor Mountain Wild Horse Range. Thesis, University of Michigan, Ann Arbor. 130p.

Feist, J.D. and D.R. McCullough. 1975. Reproduction in feral horses. *J. Reprod. Fert., Suppl.* 23:13-18.

Feist, J.D. and D.R. McCullough. 1976. Behavior patterns and communication in feral horses. *Z. Tierpsychol.* 41:337-371.

Fiske, J.C. 1979. *How horses learn*. S. Greene Press, Brattleboro, Vermont. 148p.

Fiske, J.C. and G.D. Potter. 1979. Discrimination reversal learning in yearling horses. *J. Anim. Sci.* 49:583-588.

Flade, J.E. 1958. Die Verteilung der Geburten bei Pferden auf die Tageszeit. *Tierzucht* 12:93-95.

Ford, B. and R.R. Keiper. 1979. *The island ponies: An environmental study of their life on Assateague*. W. Morrow, New York. 95p.

Fox, M.W. 1965. *Canine behavior*. C.C. Thomas, Springfield, Illinois.

Francis-Smith, K. 1978. The nursing behaviour of foals. *Proc. 1st World Congr. Ethol. Appl. to Zootechnics, Madrid, Spain*. Vol. II(E-4-01):97.

Fraser, A.C. 1967. Restraint in the horse. *Vet. Rec.* 80: 56-64.

Fraser, A.F. 1970. Some observations on equine oestrus. *Brit. Vet. J.* 126:656-657.

Fraser, A.F., H. Hastie, R.B. Callicott and S. Brownlie. 1975. An exploratory ultrasonic study on quantitative fetal kinesis in the horse. *Appl. Anim. Ethol.* 1:395-404.

Fraser, J.A. 1969. Some observations on the behaviour of the horse in pain. *Brit. Vet. J.* 125:150-151.

Fretz, P.B. 1977. Behavioral virilization in a brood mare. *Appl. Anim. Ethol.* 3:277-280.

Gabel, A.A. 1972. Drugs used in restraint. Pages 655-659 in E.J. Catcott and J.F. Smithcors, eds. *Equine medicine and surgery.* 2nd edition. American Veterinary Publications, Wheaton, Illinois.

Gabel, A.A. and E.W. Jones. 1972. General anesthesia. Pages 659-664 in E.J. Catcott and J.F. Smithcors, eds. *Equine medicine and surgery.* 2nd edition. American Veterinary Publications, Wheaton, Illinois.

Gardner, L.P. 1933. The responses of horses to the situation of a closed feed box. *J. Comp. Psychol.* 15:445-467.

Gardner, L.P. 1937a. The responses of horses in a discrimination problem *J. Comp. Psychol.* 23:13-34.

Gardner, L.P. 1937b. Responses of horses to the same signal in different positions. *J. Comp. Psychol.* 23:305-332.

Gardner, L.P. 1942. Conditioning horses and cows to the pail as a signal. *J. Comp. Psychol.* 34:29-41.

Gates, S. 1979. A study of the home ranges of free-ranging Exmoor ponies. *Mammal Rev.* 9:3-18.

Giebel, H.-D. 1958. Visuelles Lernvermögen bei Einhufern. *Zool. Jb.* 67:487-520.

Ginther, O.J. 1974. Occurrence of anestrus, estrus, diestrus, and ovulation over a 12-month period in mares. *Amer. J. Vet. Res.* 35:1173-1179.

Ginther, O.J. 1979. *Reproductive biology of the mare: basic and applied aspects.* O.J. Ginther, Cross Plains, Wisconsin. 413p.

Ginther, O.J. and P.E. Meckley. 1972. Effect of intrauterine infusion on length of diestrus in cows and mares. *Vet. Med.* 67:751-754.

Ginther, O.J., H.L. Whitmore and E.L. Squires. 1972. Characteristics of estrus, diestrus, and ovulation in mares and effects of season and nursing. *Amer. J. Vet. Res.* 33:1935-1939.

Glendinning, S.A. 1977. The behaviour of sucking foals. *Brit. Vet. J.* 133:192.

Göbel, F. and K. Zeeb. 1963. Primitivpferde und ihre Haltung. *Tierärztl. Umsch.* 18:64, 67-71.

Goldschmidt-Rothschild, B. von and B. Tschanz. 1978. Soziale organisation und verhalten einer Jungtierherde beim Camargue-Pferd. *Z. Tierpsychol.* 46:372-400.

Green, N.F. and H.D. Green. 1977. The wild horse population of Stone Cabin Valley, Nevada: a preliminary report. Pages 59-65 in *Proc. National Wild Horse Forum, 4-7 April 1977.* R-127, Cooperative Extension Service, University of Nevada, Reno.

Grogan, J.W. 1951. The gaits of horses. *J. Amer. Vet. Med. Assn.* 119:112-117.

Grzimek, B. 1943a. Bergrüssung zweier Pferde: das Erkennen von Phantomen und Bildern. *Z. Tierpsychol.* 5:465-480.

Grzimek, B. 1943b. Heimfinde-Versuche mit Pferden. *Z. Tierpsychol.* 5:481-497.

Grzimek, B. 1944a. Scheuversuche mit Pferden. *Z. Tierpsychol.* 6:26-40.

Grzimek, B. 1944b. Das Erkennen von Menschen durch Pferde. *Z. Tierpsychol.* 6:110-120.

Grzimek, B. 1949a. Ein Fohlen, des kein Pferd kannte. *Z. Tierpsychol.* 6:391-405.

Grzimek, B. 1949b. Die Rechts- und Linkshandigkeit bei Pferden, Papageien und Affen. *Z. Tierpsychol.* 6:406-432.

Grzimek, B. 1949c. Gedächtnisversuche mit Pferden. *Z. Tierpsychol.* 6:445-454.

Grzimek, B. 1949d. Rangordnungsversuche mit Pferden. *Z. Tierpsychol.* 6:455-464.

Grzimek, B. 1952. Versuche über das Farbsehen von Pflanzenessern. I. Das farbige Sehen (und die Sehschärfe) von Pferden. *Z. Tierpsychol.* 9:23-39.

Haag, E.L., R. Rudman and K.A. Houpt. 1980. Avoidance, maze learning and social dominance in ponies. *J. Anim. Sci.* 50:329-335.

Haenlein, G.F.W., R.D. Holdren and Y.M. Yoon. 1966. Comparative response of horses and sheep to different physical forms of alfalfa hay. *J. Anim. Sci.* 25:740-743.

Hafez, E.S.E., M. Williams and S. Wierzbowski. 1962. The behaviour of horses. Pages 370-396 in E.S.E. Hafez, ed. *The behaviour of domestic animals.* 1st edition. Williams and Wilkins, Baltimore.

Hamilton, G.V. 1911. A study of trial and error reactions in mammals. *J. Anim. Behav.* 1:33-66.

Hansen, R.M. 1976. Foods of free-ranging horses in southern New Mexico. *J. Range Mgmt.* 29:347.

Harlow, H.F. 1949. The formation of learning sets. *Psychol. Rev.* 56:51-56.

Hassenberg, L. 1971. *Verhalten bei Einhufern.* Neue Brehm-Bücherei 427, Ziemsen, Wittenberg Lutherstadt. 159p.

Hechler, B. 1971. Beitrag zur Ethologie des Islandpferdes. Inaugural-Dissertation, Universität Giessen. 90p.

Heglund, N.C., C.R. Taylor and T.A. McMahon. 1974. Scaling stride frequency and gait to animal size: mice to horses. *Science* 186:1112-1113.

Heird, J.C., A.M. Lennon and R.W. Bell. 1981. Effects of early experience on the learning ability of yearling horses. *J. Anim. Sci.* 53:1204-1209.

Hendrikse, J. 1972. Draagtijden van Nederlandse paarderassen. *Tijdschr. Diergeneesk.* 97:477-480.

Hendrix, G. 1968. Unverbalized awareness as an agency for transfer of learning. Paper presented at Symposium No. 40. The effects of conscious purpose on human adaptation. Wenner-Gren Foundation for Anthropological Research, Burg Wartenstein, Austria.

Heptner, V.G., A.A. Nasimovič and A.G. Bannikov. 1966. Paarhufer und Unpaarhufer. Band I in V.G. Heptner and N.P. Naumov, ed. *Die Säugetiere der Sowjetunion.* G. Fischer, Jena.

Hildebrand, M. 1959. Motions of the running cheetah and horse. *J. Mammal.* 40:481-495.

Hildebrand, M. 1965. Symmetrical gaits of horses. *Science* 150:701-708.

Hildebrand, M. 1977. Analysis of asymmetrical gaits. *J. Mammal.* 58:131-156.

Hintz, H.F., R.L. Hintz, D.H. Lein and L.D. Van Vleck. 1979. Length of gestation periods in Thoroughbred mares. *J. Equine Med. Surg.* 3:289-292.

Houpt, K.A. 1981. Equine behavior problems in relation to humane management. *Int. J. Stud. Anim. Prob.* 2:329-337.

Houpt, K.A. and H.F. Hintz. 1981. Changes in the mare-foal bond with time. Paper presented at Animal Behavior Society Annual Meeting, University of Tennessee, Knoxville.

Houpt, K.A., K. Law and V. Martinisi. 1978. Dominance hierarchies in domestic horses. *Appl. Anim. Ethol.* 4:273-283.

Howell, C.E. and W.C. Rollins. 1951. Environmental sources of variation in the gestation length of the horse. J. Anim. Sci. 10:789-796.

Hoyt, D.F. and C.R. Taylor. 1981. Gait and the energetics of locomotion in horses. *Nature* 292:239-240.

Hubbard, R.E. and R.M. Hansen. 1976. Diets of wild horses, cattle and mule deer in the Piceance Basin, Colorado. *J. Range Mgmt.* 29:389-392.

Hughes, A. 1977. The topography of vision in mammals of contrasting life style: comparative optics and retinal organization. Pages 613-756 in F. Crescitelli, ed. *The visual system in vertebrates. Vol. 7/5. Handbook of Sensory Physiology.* Springer-Verlag, New York.

Hughes, J.P. 1978. Inducing springtime estrus in the mare. *J. Amer. Vet. Med. Assn.* 172:817.

Hughes, J.P., G.H. Stabenfeldt and J.W. Evans. 1972a. Clinical and endocrine aspects of the estrous cycle of the mare. *Proc. 18th Annual Conv. Amer. Assn. Equine Prac.* 119-148.

Hughes, J.P., G.H. Stabenfeldt and J.W. Evans. 1972b. Estrous cycle and ovulation in the mare. *J. Amer. Vet. Med. Assn.* 161:1367-1374.

Hughes, R.D., P. Duncan and J. Dawson. 1981. Interactions between Camargue horses and horseflies (Diptera: Tabanidae). *Bull. Ent. Res.* 71:227-242.

Hurtgen, J.P. and H.L. Whitmore. 1979. Induction of estrus and ovulation by endometrial biopsy in mares with prolonged diestrus. *J. Amer. Vet. Med. Assn.* 175:1196-1197.

Irvine, D.S., B.R. Downey, W.G. Parker and J.J. Sullivan. 1975. Duration of oestrus and time of ovulation in mares treated with synthetic Gn-RH (AY-24,031). *J. Reprod. Fert., Suppl.* 23:279-283.

Janis, C. 1976. The evolutionary strategy of the Equidae and the origins of rumen and cecal digestion. *Evolution* 30:757-774.

Janzen, D.H. 1978. How do horses find their way home? *Biotropica* 10:240.

Jaworowska, M. 1976. Verhaltensbeobachtungen an primitiven polnischen Pferden, die in einem polnischen Wald-Schutzgebiet--in Freiheit Lebend--erhalten werden. *Säugetierk. Mitt.* 24:241-268.

Jeffcott, L.B. 1972. Observations on parturition in crossbred pony mares. *Equine Vet. J.* 4:209-213.

Kare, M. 1971. Comparative study of taste. Pages 278-292 in L.M. Beidler, ed. *Chemical senses: taste.* Vol. 4/2. *Handbook of sensory physiology.* Springer-Verlag, New York.

Keiper, R.R. 1975. The behavior, ecology and social organization of the feral ponies of Assateague Island. Report submitted to Assateague Island National Seashore, National Park Service, Maryland.

Keiper, R.R. 1976a. Social organization of feral ponies. *Proc. Pennsylvania Acad. Sci.* 50:69-70.

Keiper, R.R. 1976b. Interactions between cattle egrets and feral ponies. *Proc. Pennsylvania Acad. Sci.* 50:89-90.

Keiper, R.R. 1979a. Anti-insect behaviors of feral ponies. Paper presented at Animal Behavior Society Annual Meeting, Tulane University, New Orleans.

Keiper, R.R. 1979b. Population dynamics of feral ponies. Paper presented at Symposium on the Ecology and Behavior of Wild and Feral Equids, University of Wyoming, Laramie.

Keiper, R.R. 1980. Band fission and formation in feral ponies. Paper presented at Animal Behavior Society Annual Meeting, Colorado State University, Fort Collins.

Keiper, R.R. and M.A. Keenan. 1980. Nocturnal activity patterns of feral ponies. *J. Mammal.* 61:116-118.

Kiley, M. 1972. The vocalizations of ungulates, their causation and function. *Z. Tierpsychol.* 31:171-222.

Kiley-Worthington, M. 1976. The tail movements of ungulates, canids and felids, with particular reference to their causation and function as displays. *Behaviour* 56:69-115.

Klingel, H. 1977. Communication in Perrissodactyla. Pages 715-727 in T.A. Sebeok, ed. *How animals communicate.* Indiana University Press, Bloomington.

Knight, H.D. 1972. Other bacterial infections. Pages 85-113 in E.J. Catcott and J.F. Smithcors, eds. *Equine medicine and surgery.* 2nd edition. American Veterinary Publications, Wheaton, Illinois.

Knill, L.M., R.D. Eagleton and E. Harver. 1977. Physical optics of the equine eye. *Amer. J. Vet. Res.* 38:735-737.

Koch, W. 1951. Psychogene Beeinflussung des Geburtstermins bei Pferden. *Z. Tierpsychol.* 8:441-443.

Koegel, A. 1954. Vom Öffnen von Türen durch Tiere. *Z. Tierpsychol.* 11:495-496.

Kosiniak, K. 1975. Characteristics of the successive jets of ejaculated semen of stallions. *J. Reprod. Fert., Suppl.* 23:59-61.

Kownacki, M., E. Sasimowski, M. Budzyński, T. Jezierski, M. Kapron, B. Jelen, M. Jaworska, R. Dziedzic, A. Seweryn, Z. Solmka. 1978. Observations of the twenty-four hours rhythm of natural behaviour of Polish primitive horse bred for conservation of genetic resources in a forest reserve. *Genetica Polonica* 19:61-77.

Kratzer, D.D., W.M. Netherland, R.E. Pulse and J.P. Baker. 1977. Maze learning in Quarter Horses. *J. Anim. Sci.* 45:896-902.

Lawson, A.C. 1908. *The California earthquake of April 18, 1906.* Report of the State Earthquake Investigation Committee. Publ. 87. Carnegie Institution, Washington, D.C.

Leahy, J.R. and P. Barrow. 1953. *Restraint of animals.* 2nd edition. Cornell Campus Store, Ithaca, New York. 269p.

Leblanc, M.A. and M.F. Bouissou. 1981. Mise au point d'une épreuve destinée à l'étude de la reconnaissance du jeune par la mère chez le cheval. *Biol. Behav.* 6:283-290.

Littlejohn, A. 1970. Behaviour of horses recovering from anaesthesia. *Brit. Vet. J.* 127:617-621.

Littlejohn, A. and R. Munro. 1972. Equine recumbency. *Vet. Rec.* 90:83-85.

Loy, R.G. 1967. How the photoperiod affects reproductive activity in mares. *Mod. Vet. Pract.* 48(5):47-49.

Loy, R.G. and J.P. Hughes. 1966. The effects of human chorionic gonadotrophin on ovulation. length of estrus, and fertility in the mare. *Cornell Vet.* 56:41-50.

Loy, R.G., J.P. Hughes, W.P.C. Richards and S.M. Swan. 1975. Effects of progesterone on reproductive function in mares after parturition. *J. Reprod. Fert. Suppl.* 23:291-295.

Loy, R.G. and S.M. Swan. 1966. Effect of exogenous progestogens on reproductive phenomena in mares. *J. Anim. Sci.* 25:821-826.

McPheeters, G.M., Jr. 1973. Dominance hierarchies in horses for the behaviors of drinking, grazing, and resting. Graduate Research Paper, Southern Illinois University, Carbondale. 37p.

Maday, S. von. 1912. *Psychologie des Pferdes und der Dressur.* P. Parey Verlag, Berlin. 349p.

Mader, D.R. and E.O. Price. 1980. Discrimination learning in horses: effects of breed, age and social dominance. *J. Anim. Sci.* 50:962-965.

Magne de la Croix, P. 1936. The evolution of locomotion in mammals. *J. Mammal.* 17:51-54.

Marinier, S. 1980. Selective grazing behaviour in horses. Dissertation, University of Natal, Durban, South Africa. 235p.

Martin-Rosset, W., M. Doreau and J. Cloix. 1978. Etude des activités d'un troupeau de poulinières de trait et de leurs poulains au pâturage. *Ann. Zootech.* 27:33-45.

Marwick, C. 1967. Towards a more "human" horse. *New Sci.* 33(529):76.

Matthews, R.G., R.T. Ropiha and R.M. Butterfield. 1967. The phenomenon of foal heat in mares. *Aust. Vet. J.* 43:579-582.

Miller, R. 1979. Social organization and movements of feral horses in Wyoming's Red Desert. Paper presented at Symposium on the Ecology and Behavior of Wild and Feral Equids, University of Wyoming, Laramie.

Miller, R. 1980. Band organization and stability in Red Desert feral horses. Pages 113-128 in R.H. Denniston, ed. *Symposium on the ecology and behavior of wild and feral equids, 6-8 September 1979.* University of Wyoming, Laramie.

Miller, R. and R.H. Denniston. 1979. Interband dominance in feral horses. *Z. Tierpsychol.* 51:41-47.

Miller, R.M. 1966. Psychological effects of succinylcholine chloride immobilization on the horse. *Vet. Med./Sm. Anim. Clin.* 61:941-944.

Mintscheff, P. and R. Prachoff. 1960. Versuche zur Erhöhung der Befruchtungsfahigkeit der Stute durch Hungerdiät während der Brunst. *Zuchthyg. Fortpflstör. Besam. Haustiere* 4:40-48.

Montgomery, G.G. 1957. Some aspects of the sociality of the domestic horse. *Trans. Kansas Acad. Sci.* 60:419-424.

Müller, W. 1942. Schreit das Pferd bei grossen Schmerzen und in Todesnot? *Z. Veterinärk.* 54:273-278.

Munaretto, K.R. 1980. Reciprocal voice recognition from auditory cues by a mare and her foal *(Equus caballus)*. Unpublished Research Report, Southern Illinois University Carbondale.

Myers, R.D. and D.C. Mesker. 1960. Operant responding in a horse under several schedules of reinforcement. *J. Exper. Analyt. Behav.* 3:161-164.

Neely, D.P., J.P. Hughes, G.H. Stabenfeldt and J.W. Evans. 1975. The influence of intrauterine saline infusion on luteal function and cyclical activity in the mare. *J. Reprod. Fert., Suppl.* 23:235-239.

Nickel, R., A. Schummer and E. Seiferle. 1960. *Lehrbuch der Anatomie der Haustiere.* P. Parey, Berlin.

Nicolas, E. 1930. *Veterinary and comparative ophthalmology.* H. & W. Brown, London.

Nishikawa, Y. 1954. Strength of sexual desire and properties of ejaculate of horse after castration. *Bull. Nat. Inst. Agr. Sci.,* Japan, Ser. G., No. 8, 161-167.

Nishikawa, Y. 1959. *Studies on reproduction in horses.* Japan Racing Assn. Tokyo. 340p.

Nobbe, D.E. 1974. Instrumental conditioning of a horse using grain as a reinforcer. Unpublished Research Report, Southern Illinois University, Carbondale.

Nobbe, D.E. 1978. Delayed response in the horse. Paper presented at Animal Behavior Society Midwestern Conference, West Lafayette, Indiana.

Nobis, G. 1971. *Vom Wildpferd zum Hauspferd. Studien zur Phylogenie pleistozäner Equiden Eurasiens und das Domestikations-problem unserer Hauspferde.* Böhlau Verlag, Köln. 96p.

Noden, P.A., W.D. Oxender and H.D. Hafs. 1975. The cycle of oestrus, ovulation and plasma levels of hormones in the mare. *J. Reprod. Fert., Suppl.* 23:189-192.

Ödberg, F.O. 1969. Bijdrage tot de studie van de gedragingen van het paard *(E. caballus* L.). Thesis, State University Ghent, Belgium. 165p.

Ödberg, F.O. 1973. An interpretation of pawing by the horse *(Equus caballus* Linnaeus), displacement activity and original functions. *Säugetierk. Mitt.* 21:1-12.

Ödberg, F.O. 1978. A study of the hearing ability of horses. *Equine Vet. J.* 10:82-84.

Ödberg, F.O. and K. Francis-Smith. 1976. A study on eliminative and grazing behavior--the use of the field by captive horses. *Equine Vet. J.* 8:147-149.

Ödberg, F.O. and K. Francis-Smith. 1977. Studies on the formation of ungrazed eliminative areas in fields used by horses. *Appl. Anim. Ethol.* 3:27-34.

Olberg, G. 1959. Der artfremde Kumpan in der Umwelt des Pferdes. *Kosmos* 55:431-436.

Olsen, F.W. and R.M. Hansen. 1977. Food relations of wild free-roaming horses to livestock and big game, Red Desert, Wyoming. *J. Range Mgmt.* 30:17-20.

Oxender, W.D., P.A. Noden and H.D. Hafs. 1975. Oestrus, ovulation and plasma hormones after prostaglandin $F_{2\alpha}$ in mares. *J. Reprod. Fert., Suppl.* 23:251-255.

Pellegrini, S. 1971. Home range, territoriality, and movement patterns of wild horses in the Wassuk Range of western Nevada. Thesis, University of Nevada, Reno. 39p.

Penick, J., Jr. 1976. *The New Madrid earthquakes of 1811-1812.* University of Missouri Press, Columbia. 181p.

Pfaffenberger, C.J. and J.P. Scott. 1959. The relationship between delayed socialization and trainability in guide dogs. *J. Genet. Psychol.* 95:145-155.

Pfungst, O. 1907. *Das Pferd des Herrn von Osten (Der "kluge Hans"). Ein Beitrag zur experimentellen Tier- und Menschenpsychologie.* Barth, Leipzig.

Pickett, B.W. 1974. Evaluation of stallion semen. Pages 60-87 in O.R. Adams, *Lameness in horses.* 3rd edition. Lea and Febiger, Philadelphia.

Pickett, B.W., L.C. Faulkner and T.M. Sutherland. 1970. Effect of month and stallion on seminal characteristics and sexual behavior. *J. Anim. Sci.* 31:713-728.

Pickett, B.W., L.C. Faulkner, G.E. Seidel Jr., W.E. Berndtson and J.L. Voss. 1976. Reproductive physiology of the stallion. Part 6. Seminal and behavioral characteristics. *J. Anim. Sci.* 43:617-625.

Pickett, B.W., J.L. Voss and E.L. Squires. 1977. Impotence and abnormal sexual behavior in the stallion. *Theriogenology* 8:329-347.

Pineda, M.H. and O.J. Ginther. 1972. Inhibition of estrus and ovulation in mares treated with an antiserum against an equine pituitary fraction. *Amer. J. Vet. Res.* 33:1775-1780.

Pisa, A. 1939. Über den binokularen Gesichtsraum bei Haustieren. *v. Graefe's Arch. Ophthal.* 140:1-54.

Podliachouk, L. and M. Kaminski. 1971. Comparative investigations of Equidae. A study of blood groups and serum proteins in a sample of *Equus prezwalskii* Poliakoff. *Anim. Blood Groups and Biochem. Genetics* 2:239-242.

Popov, N.F. 1956. Features of higher nervous activity of horses. [In Russian] *Zh. v. n. Deiatel'* 6:718-725.

Prahov, R. 1959. Inducing oestrus and accelerating ovulation in mares by reflex methods. [In Russian] *Nauč. Trud. naučnoizsled. Inst. Razvǎd. Bol. izkustv. Osemen. selskostop. Životn.* 1:69-76. (*Anim. Breed. Abstr.* 30:157.)

Prince J.H. 1970. The eye and vision. Pages 1135-1159 in M.J. Swenson, ed. *Duke's physiology of domestic animals.* 8th edition. Cornell University Press, Ithaca, New York.

Prince, J.H., C.D. Diesem, I. Eglitis and G.L. Ruskell. 1960. *Anatomy and histology of the eye and orbit in domestic animals.* C.C. Thomas, Springfield, Illinois. 307p.

Radinsky, L.B. 1966. The adaptive radiation of the phenacodontid condylarths and the origin of the Perissodactyla. *Evolution* 20:408-417.

Ralston, S.L., G. Van den Broek and C.A. Baile. 1979. Feed intake patterns and associated blood glucose, free fatty acid and insulin changes in ponies. *J. Anim. Sci.* 49:838-845.

Randall, R.P., W.A. Schurg and D.C. Church. 1978. Response of horses to sweet, salty, sour and bitter solutions. *J. Anim. Sci.* 47:51-55.

Rasbech, N.O. 1975. Ejaculatory disorders of the stallion. *J. Reprod. Fert., Suppl.* 23:123-128.

Reed, L.C. 1980. Onset and sequential development of perinatal behaviors in American Saddlebred foals. Graduate Research Paper, Southern Illinois University, Carbondale. 46p.

Rhine, J.B. and L.E. Rhine. 1929a. An investigation of a "mind-reading" horse. *J. Abnorm. & Soc. Psychol.* 23:449-466.

Rhine, J.B. and L.E. Rhine. 1929b. Second report on Lady, the "mind-reading" horse. *J. Abnorm. & Soc. Psychol.* 24:287-292.

Rooney, J.R. 1971. *Clinical neurology of the horse.* KNA Press, Kennett Square, Pennsylvania. 104p.

Rooney, J.R. 1973. Two cervical reflexes in the horse. *J. Amer. Vet. Med. Assn.* 162:117-118.

Rooney, J.R. 1978. The role of the neck in locomotion. *Mod. Vet. Prac.* 59:211-213.

Rooney, J.R. 1979. Postural reflexes in foals. *Mod. Vet. Prac.* 60:392-394.

Rooney, J.R. 1981. *The mechanics of the horse.* R.E. Krieger, Huntington, New York. 92p.

Ropiha, R.T., R.G. Matthews, R.M. Butterfield, F.P. Moss and W.J. McFadden. 1969. The duration of pregnancy in Thoroughbred mares. *Vet. Rec.* 84:552-555.

Rossdale, P.D. 1967a. Clinical studies on the newborn Thoroughbred foal. I. Perinatal behaviour. *Brit. Vet. J.* 123:470-481.

Rossdale, P.D. 1967b. Clinical studies on the newborn Thoroughbred foal. II. Heart rate, auscultation and electrocardiogram. *Brit. Vet. J.* 123:521-532.

Rossdale, P.D. 1968a. Clinical studies on the newborn Thoroughbred foal. III. Thermal stability. *Brit. Vet. J.* 124:18-22.

Rossdale, P.D. 1968b. Perinatal behavior in the Thoroughbred horse. Pages 227-237 in M.W. Fox, ed. *Abnormal behavior of domestic animals.* C.C. Thomas, Springfield, Illinois.

Rossdale, P.D. 1969. Measurements of pulmonary ventilation in normal newborn Thoroughbred foals during the first three days of life. *Brit. Vet. J.* 125:157-162.

Rossdale, P.D. and L.W. Mahaffey. 1958. Parturition in the Thoroughbred mare with particular reference to blood deprivation in the new-born. *Vet. Rec.* 70:142-152.

Rossdale, P.D. and S.W. Ricketts. 1974. *The practice of equine stud medicine.* Baillière Tindall, London. 421p.

Rossdale, P.D. and R.V. Short. 1967. The time of the foaling of Thoroughbred mares. *J. Reprod. Fert.* 13:341-343.

Rubenstein, D.I. 1978. Islands and their effects on the social organization of feral horses. Paper presented at Symposium on Social Behavior on Islands, Animal Behavior Society Annual Meeting, University of Washington, Seattle.

Rubenstein, D.I. 1981. Behavioural ecology of island feral horses. *Equine Vet. J.* 13:27-34.

Rubin, L., C. Oppegard and H.F. Hintz. 1980. The effect of varying the temporal distribution of conditioning trials on equine learning behavior. *J. Anim. Sci.* 50:1184-1187.

Ruckebusch, Y. 1970. Un problème controversé: la perte de vigilance chez le cheval et la vache au cours du sommeil. *Les Cahiers de Med. Vet.* 39:210-225.

Ruckebusch, Y. 1972. The relevance of drowsiness in the circadian cycle of farm animals. *Anim. Behav.* 20:637-643.

Ruckebusch, Y., P. Barbey and P. Guillemot. 1970. Les états de sommeil chez le cheval *(Equus caballus). C.R. Seances Soc. Biol. Filiales* 164:658-665.

Rudman, R., E. Haag and K.A. Houpt. 1980. Avoidance, maze learning, and social dominance in ponies. Pages 99-111 in R.H. Denniston, ed. *Symposium on the ecology and behavior of wild and feral equids, 6-8 September 1979.* University of Wyoming, Laramie.

Ryder, O.A. 1977. The chromosomes of a rare equine: *Equus hemionus kulan. ZooNooz* 50(12): 15.

Ryder, O.A., N.C. Epel and K. Benirschke. 1978. Chromosome banding studies of the Equidae. *Cytogenet. Cell Genet.* 20:323-350.

Salter, R.E. 1978. Ecology of feral horses in western Alberta. Thesis, University of Alberta, Edmonton. 233p.

Salter, R.E. and R.J. Hudson. 1979. Feeding ecology of feral horses in western Alberta. *J. Range Mgmt.* 32:221-225.

Salter, R.E. and D.J. Pluth. 1980. Determinants of mineral lick utilization by feral horses. *Northwest Sci.* 54:109-118.

Sambraus, H.H. and D. Sambraus. 1975. Prägung von Nutztieren auf Menschen. *Z. Tierpsychol.* 38:1-17.

Schäfer, M. 1975. *Language of the horse.* Arco, New York. 187p.

Schäfer, M. 1978. Pferd. Pages 214-248 in H.H. Sambraus, ed. *Nutztierethologie.* P. Parey, Berlin.

Schaffer, J. 1940. *Die Hautdrüsenorgane der Säugetiere.* Urban und Schwarzenberg, Berlin. 464p.

Schneider, K.M. 1930. Das Flehmen. *Zool. Gart., Lpz.* 3:183-198.

Schneider, K.M. 1931. Das Flehmen. II. *Zool. Gart., Lpz.* 4:349-364.

Schneider, K.M. 1932a. Das Flehmen. III. *Zool. Gart., Lpz.* 5:200-226.

Schneider, K.M. 1932b. Das Flehmen. IV. *Zool. Gart., Lpz.* 5:287-297.

Schneider, K.M. 1934. Das Flehmen, V. Zur psychologischen Deutung. *Zool. Gart., Lpz.* 7:182-201.

Schoen, A.M.S., E.M. Banks and S.E. Curtis. 1976. Behavior of young Shetland and Welsh ponies *(Equus caballus). Biol. Behav.* 1:199-216.

Seiferle, E. 1960. Schmerz und Angst bei Tier und Mensch. *Deutsche Tierärztl. Wochenschr.* 67:275-278, 332-334.

Seunig, W. 1956. *Horsemanship.* Doubleday, Garden City, New York. 390p.

Sharp, D.C., L. Kooistra and O.J. Ginther. 1975. Effects of artificial light on the oestrous cycle of the mare. *J. Reprod. Fert., Suppl.* 23:241-246.

Siegmund, O.H. ed. 1973. *The Merck veterinary manual.* Merck, Rahway, New Jersey, 1600p.

Simpson, G.G. 1951. *Horses.* Oxford University Press, New York. 247p.

Sisson, S. and J.D. Grossman. 1953. *The anatomy of the domestic animals.* Saunders, Philadelphia. 972p.

Sivak, J.G. and D.B. Allen. 1975. An evaluation of the ramp retina on the horse eye. *Vision Res.* 15:1353-1356.

Skorkowski, E. 1956. Systematik und Abstammung des Pferdes. *Z. Tierzüchtg. Züchtungsbiol.* 68:42-74.

Skorkowski, E. 1971. Ewolucja gatunku konia. *Przegl. Zool.* 15:308-315.

Soemmerring, D.W. 1818. De oculorum hominis animaliumque. Dissertation, University of Göttingen, Göttingen.

Sondaar, P.Y. 1968. The osteology of the manus of fossil and recent Equidae, with special reference to phylogeny and function. *Verh. Konink. Nederl. Acad. Wetens.* 25(1).

Sondaar, P.Y. 1969. Some remarks on horse evolution and classification. *Z. Säugetierk.* 34:307-311.

Spector, W.S. ed. 1956. *Handbook of biological data.* W.B. Saunders, Philadelphia. 584p.

Speed, J.G. and M. Etherington. 1952a. An aspect of the evolution of British horses. *Brit. Vet. J.* 108:147-153.

Speed, J.G. and M.G. Etherington. 1952b. The Exmoor Pony--and a survey of the evolution of horses in Britain. Part I. *Brit. Vet. J.* 108:329-338.

Speed, J.G. and M.G. Etherington. 1953. The Exmoor Pony--and a survey of the evolution of horses in Britain. Part II. The Celtic Pony. *Brit. Vet. J.* 109:315-320.

Stabenfeldt, G.H. and J.P. Hughes. 1977. Reproduction in horses. Pages 401-431 in H.H. Cole and P.T. Cupps, eds. *Reproduction in domestic animals.* 3rd edition. Academic Press, New York.

Stabenfeldt, G.H., J.P. Hughes, J.W. Evans and I.I. Geschwind. 1975. Unique aspects of the reproductive cycle of the mare. *J. Reprod. Fert., Suppl.* 23:155-160.

Stanley, W.C. 1965. The passive person as a reinforcer in isolated beagle puppies. *Psychon. Sci.* 2:21-22.

Stanley, W.C. and O. Elliot. 1962. Differential human handling as reinforcing events and as treatments influencing later social behavior in Basenji puppies. *Psychol. Rep.* 10:775-788.

Stebbins, M.C. 1974. Social organization in free-ranging Appaloosa Horses. Thesis, Idaho State University, Pocatello. 306p.

Steinhart, P. 1937. Der Schlaf des Pferdes. *Z. Veterinärk.* 49:145-157, 193-232.

Stevenson, S.M. 1975. The expressive movements accompanying the sound emissions of the horse. Thesis, Southern Illinois University, Carbondale. 81p.

Stowe, H.D. 1969. Composition of a complete purified equine diet. *J. Nutr.* 98:330-334.

Studiencow, A. 1953. *Veterinary obstetrics and gynaecology.* [In Russian] Gos. Izd. Sielsk. Lit., Moscow.

Talukdar, A.H., M.L. Calhoun and A.W. Stinson. 1970. Sensory end organs in the upper lip of the horse. *Amer. J. Vet. Res.* 31:1751-1754.

Talukdar, A.H., M.L. Calhoun and A.W. Stinson. 1972. Microscopic anatomy of the skin of the horse. *Amer. J. Vet. Res.* 33:2365-2390.

Taylor, E.L. 1954. Grazing behaviour and helminthic disease. *Brit. J. Anim. Behav.* 2:61-62.

Temple, J.L. 1963. Vices. Pages 787-792 in J.F. Bone et al., eds. *Equine medicine and surgery.* American Veterinary Publications, Wheaton, Illinois.

Tricker, R.A.R. and B.J.K. Tricker. 1967. *The science of movement.* American Elsevier, New York. 284p.

Trotter, G.W. and W.A. Aanes. 1981. A complication of cryptorchid castration in three horses. *J. Amer. Vet. Med. Assn.* 179:246-248.

Trum, B.F. 1950. The estrous cycle of the mare. *Cornell Vet.* 40:17-23.

Trumler, E. 1959. Das "Rossigkeitsgesicht" und ähnliches Ausdrucksverhalten bei Einhufern. *Z. Tierpsychol.* 16:478-488.

Tschanz, B. 1979. Sozialverhalten beim Camarguepferd--Dokumentierverhalten bei Hengsten (Freilandaufnahmen). *Publ. Wiss. Film,* Göttingen, Sekt. Biol., Ser. 12, Nr. 12/D1284. 16p.

Tschanz, B. 1980. Sozialverhalten beim Camarguepferd--Paarungsverhalten und Herdenstruktur (Freilandaufnahmen). *Publ. Wiss. Film.,* Göttingen, Sekt. Biol., Ser. 13, Nr. 34/D1318. 15p.

Turner, J.W. Jr., A. Perkins, E. Gevers and J.F. Kirkpatrick. 1979. Quantitative aspects of elimination behavior in feral stallions. Paper presented at Symposium on the Ecology and Behavior of Wild and Feral Equids, University of Wyoming, Laramie.

Turner, J.W. Jr., A. Perkins and J.F. Kirkpatrick. 1981. Elimination marking behavior in feral horses. *Can. J. Zool.* 59:1561-1566.

Tyler, S.J. 1969. The behaviour and social organisation of the New Forest ponies. Dissertation, University of Cambridge, Cambridge. 188p.

Tyler, S.J. 1972. The behaviour and social organization of the New Forest ponies. *Anim. Behav. Monogr.* 5:85-196.

Vandeplassche, M. 1955. Ejakulationsstörungen beim Hengst. *Fortpfl. Zuchthyg. Haustierbesam.* 5:134-137.

Vaughan, J.T. 1972. Physical restraint. Pages 690-709 in E.J. Catcott and J.F. Smithcors, eds. *Equine medicine and surgery.* 2nd edition. American Veterinary Publications, Wheaton, Illinois.

Veeckman, J. 1979. Aberrant sexual behaviour of a covering stallion and its ethological solution. *Appl. Anim. Ethol.* 5:292.

Veeckman, J. and F.O. Ödberg. 1978. Preliminary studies on the behavioural detection of oestrus in Belgian 'warm-blood' mares with acoustic and tactile stimuli. *Appl. Anim. Ethol.* 4:109-118.

Voith, V.L. 1975. Pattern discrimination, learning set formation, memory retention, spatial and visual reversal learning by the horse. Thesis, Ohio State University, Columbus. 32p.

Voith, V.L. 1979a. Treatment of fear-induced aggression in a horse. *Mod. Vet. Prac.* 60:835-837.

Voith, V.L. 1979b. Effects of castration on mating behavior. *Mod. Vet. Prac.* 60:1040-1041.

Voss, J.L. and B.W. Pickett. 1975. The effect of rectal palpation on the fertility of cyclic mares. *J. Reprod. Fert., Suppl.* 23:285-290.

Wallach, S.J.R. 1978. Analysis of behavioral sexual receptivity of domestic horse and pony mares *(Equus caballus)* during estrus in relation to ovulation. Dissertation, Michigan State University, East Lansing.

Walls, G.L. 1942. *The vertebrate eye.* Cranbrook Press, Bloomfield Hills, Michigan.

Walser, K. 1965. Über den Geburtsschmerz bei Tieren. *Berl. Münch. Tierärztl. Wochenschr.* 78:321-324.

Walton, A. 1960. Copulation and natural insemination. Chapter 8, vol. 1 in A.S. Parkes, ed. *Marshall's physiology of reproduction.* 3rd edition. Longmans, London.

Waring, G.H. 1970a. Perinatal behavior of foals *(Equus caballus)*. Paper presented at American Society of Mammalogists Annual Meeting, Texas A & M University, College Station.

Waring, G.H. 1970b. Primary socialization of foals *(Equus caballus)*. Paper presented at Animal Behavior Society Annual Meeting, Indiana University, Bloomington.

Waring, G.H. 1971. Sounds of the horse *(Equus caballus)*. *Bull. Ecol. Soc. Amer.* 52:45.

Waring, G.H. 1972. Socialization and behavioral development of newborn American Saddlebred Horses. Paper presented to Deutschen Veterinärmedizinischen Gesellschaft, Frieburg/Breisgau.

Waring, G.H. 1974. Behavioral adaptation of feeding in horses. *J. Anim. Sci.* 39:137.

Waring, G.H. 1978. Nursing behavior of foals *(Equus caballus)*. Paper presented at Animal Behavior Society Midwestern Conference, West Lafayette, Indiana.

Waring, G.H. and G.S. Dark. 1978. Expressive movements of horses *(Equus caballus)*. Paper presented at Animal Behavior Society Annual Meeting, University of Washington, Seattle.

Waring, G.H., S. Wierzbowski and E.S.E. Hafez. 1975. The behaviour of horses. Pages 330-369 in E.S.E. Hafez, ed. *The behaviour of domestic animals.* 3rd edition. Baillière Tindall, London.

Warnick, A.C. 1965. Reproduction and fertility in horses. Pages 103-112 in D.L. Wakeman, T.J. Cunha and J.R. Crockett. *Light horse production in Florida.* Bull. 188, Florida Department of Agriculture, Tallahassee.

Warren, J.M. and H.B. Warren. 1962. Reversal learning by horse and raccoon. *J. Genet. Psychol.* 100:215-220.

Wells, S.M. and B. von Goldschmidt-Rothschild. 1979. Social behaviour and relationships in a herd of Camargue horses. *Z. Tierpsychol.* 49:363-380.

Welsh, D.A. 1973. The life of Sable Island's wild horses. *Nature Canada* 2(2):7-14.

Welsh, D.A 1975. Population, behavioural and grazing ecology of the horses of Sable Island, Nova Scotia. Dissertation, Dalhousie University, Halifax, 403p.

Wierzbowski, S. 1958. Ejaculatory reflexes in stallions following natural stimulation and the use of the artificial vagina. *Zesz. Probl. Postep. Nauk Roln.* 11:153-156.

Wierzbowski, S. 1959. The sexual reflexes of stallions. *Roczniki Nauk Rolniczych.* 73-B-4: 753-788.

Willard, J., J.C. Willard and J.P. Baker. 1973. Dietary influence on feeding behavior in ponies. *J. Anim. Sci.* 37:227.

Willard, J.G., J.C. Willard, S.A. Wolfram and J.P. Baker. 1977. Effect of diet on cecal pH and feeding behavior of horses. *J. Anim. Sci.* 45:87-93.

Williams, M. 1957. *Horse psychology.* Countryman Press, Woodstock, Vermont. 194p.

Williams, M. 1974. Effect of artificial rearing on social behaviour of foals. *Equine Vet. J.* 6:17-18.

Winchester, C.F. 1943. Energy cost of standing in horses. *Science* 97:24.

Witherspoon, D.M. and R.B. Talbot. 1970. Nocturnal ovulation in the equine animal. *Vet. Rec.* 87:302-304.

Wolski, T.R., K.A. Houpt and R. Aronson. 1980. The role of the senses in mare-foal recognition. *Appl. Anim. Ethol.* 6:121-138.

Wright, J.G. 1943. Parturition in the mare. *J. Comp. Path. Therap.* 53:212-219.

Yeates, B.F. 1976. Discrimination learning in horses. Thesis, Texas A & M University, College Station. 31p.

Zeeb, K. 1958. Parungsverhalten von Primitivpferden in Freigehegen. *Säugetierk. Mitt.* 6:51-59.

Zeeb, K. 1959a. Das Verhalten des Pferdes bei der Auseinandersetzung mit dem Menschen. *Säugetierk. Mitt.* 7:142-192.

Zeeb, K. 1959b. Die "Unterlegenheitsgebärde" des noch nicht ausgewachsenen Pferdes *(Equus caballus)*. *Z. Tierpsychol.* 16:489-496.

Zeeb, K. 1963. *Equus caballus* (Equidae)--Erkundungs- und Meideverhalten. Encyclopaedia Cinematographica, Film No. E506. Inst. Wiss. Film, Göttingen.

Zervanos, S.M. and R.R. Keiper. 1980. Seasonal home ranges and activity patterns of feral Assateague Island ponies. Pages 3-14 in R.H. Denniston, ed. *Symposium on the ecology and behavior of wild and feral equids, 6-8 September 1979.* University of Wyoming, Laramie.

Zeuner, F.E. 1963. *A history of domesticated animals.* Hutchinson, London, 560p.

Zwolinski, J. 1966. Analysis of some phenomena connected with reproduction in mares. *Roczn. wyż. Szk. roln. Poznan.* 25:227-232. (*Anim. Breed. Abstr.* 35:391).

Index